冯＋布拉特

Report / 2005

Acknowledgements
This publication has been made possible with the help and cooperation of many individuals and institutions. Grateful acknowledgement is made to Fung + Blatt, for its inspiring work and for its kind support in the preparation of this book on Fung + Blatt for the AADCU Book Series of Contemporary Architects Studio Report In The United States.

Acknowledgements by Architect
We are very fortunate to have worked with some exceptional people through the years:

Our clients, without whom these projects would not have been possible.

Our associates -

Nicole Johnson
Elizabeth Martin
Jared Levy
Tony Pritchard
Miki Iwasaki
John Colter
Berit Eisenmann
Clarice Aguilera
Kevin Pazmino
Cathy Pack
Tres Parson
Ellen Kuch

Contractors who contributed to the projects in this book -

Bob Gornik
Lewis Sullivan
Greg Shirk
Eric Gill
Eric Chevason
Bret Goldstone
Ricky Policarpio

Our appreciation goes out to Bruce Q. Lan and the AADCU for this opportunity to catalog our work, to Haven Lin-Kirk for a crash course on the basics of book design and to Vic Liptak for copy editing the interview.
Very special thanks to Ellen Kuch for her unrelenting assistance in pulling this book together, to Tres Parson and especially to Cathy Pack, for holding down the other end of the office.

Last but not least, we thank Paulette Singley for engaging and guiding us through this critical look at where we've been.

Michael Rosner Blatt
Alice Fung

Photography Credit
Alice Fung - Jonathan Blatt Studio, Maunu deck, Yale-Maclean Residence, Schmalix Residence, Fung + Blatt Residence, Shift
Derek Rath - Dillon Street Residence
Tim Street-Porter - most of Schmalix Residence, Maunu Poolhouse
Deborah Bird - Fung + Blatt Residence, Public Storage, Shift, Kenner Studio, Campbell Apartment, Maunu Bath
Dave Lauridsen - Fung + Blatt Residence
DECO Productions - Anderson Residence
D.Vorillian - Rhodes Residence

Office of Publications
United Asia Art & Design Cooperation
www.aadcu.org
info@aadcu.org

Project Director:
Bruce Q. Lan

Coordinator:
Robin Luo

Edited and published by:
Beijing Office, United Asia Art & Design Cooperation
bj-info@aadcu.org

China Architecture & Building Press
www.china-abp.com.cn

Collaborated with:
Fung + Blatt
www.fungandblatt.com

d-Lab & International Architecture Research

School of Architecture, Central Academy of Fine Arts

Curator/Editor in Chief:
Bruce Q. Lan

Book Design:
Haven Lin-Kirk + Design studio/AADCU

ISBN: 7-112-07400-2

出版事务处
亚洲艺术与设计协作联盟／美国
www.aadcu.org
info@aadcu.org

编辑与出版：
亚洲艺术与设计协作联盟／美国
bj-info@aadcu.org

中国建筑工业出版社／北京
www.china-abp.com.cn

协同编辑：
冯＋布拉特
www.fungandblatt.com

国际建筑研究设计中心／美国

中央美术学院建筑学院／北京

主编：
蓝青

协调人：
洛宾·罗，斯坦福大学

书籍设计：
Haven Lin-Kirk ＋ 设计工作室／AADCU

Fung + Blatt

Contents

Fung + Blatt

Formed in 1990 by Alice Fung and Michael Rosner Blatt,FUNG + BLATT Architects has gained international recognition for their innovative residential design. The firm has developed a substantial body of work in residential as well as multi-family and mix-used community-based projects.
Coming from backgrounds in structural design and art, their work tests the boundary between structure, architecture, and the environmental context.
FUNG + BLATT Architects see themselves as custodians of collaborative processes, each one a unique challenge. Their expertise in structure and code issues allows the pragmatic parameters to be addressed early on, freeing them up to develop creative solutions and follow them through into thoughtful, engaging and integrated places of habitation.

艾丽丝·冯和迈克尔·罗斯纳·布拉特创建的Fung＋Blatt建筑设计事务所成立于1990年，事务所以其创新的住宅设计获得了国际上的认可。Fung 和Blatt的业务主体是住宅设计以及基于社区的多用户、多功能建筑项目。
借助于他们的结构设计和艺术背景，Fung＋Blatt的工作是在结构，建筑和周边环境之间的边界进行探索。
Fung＋Blatt把他们自己看成是合作过程中的管理者，每个过程都是一次特殊的挑战。他们在结构和业务规范方面的专业知识使得一些实际因素从一开始就能被考虑到，这使得他们能够提出创造性的解决方案，从而设计出独特的、迷人的整体居住场所。

On their approach to architecture, Fung + Blatt wrote:
Our process begins with the simultaneous investigation of several sets of concerns:
Perception + Experience
In our work, the choreography of the human experience precedes the genesis of form.
By raising questions that challenge conventional patterns, we redefine parameters for shaping space and experience; and in doing so, uncover new dimensions in the way that we relate to our physical environment.
Issues such as the inhabitants' rituals, the temporal variance in light and climate, and the tactile and acoustic qualities of materials are all explored for the ways in which they inflect upon one another in the creation of an integrated and visceral experience.
Architecture articulates spaces and activities; it also defines the transitions between them. It is in the transitions, the overlaps and the folds, that new experiences are shaped and where familiar ones are brought into new light.
Site + Structure
We believe in the responsible use of resources, and in the expression of material and structure as an integral part of how the environment is made.
The analysis of the inherent and dynamic forces of the site, and of which forces are to be enhanced or balanced, lead to the more practical determination of what is to be physically accomplished, and what the available means are to achieve it.
This inquiry develops into a structural strategy within which a formal armature is discovered, and upon which spatial and programmatic experiences are organized.
Transformation + Relevance
Good design is about how people actually inhabit and relate to their environment through time. It grows out of an understanding of the inhabitants' past, while fulfilling their current needs, and anticipating their change and growth.
A responsible and relevant environment is well suited for its immediate intended use, and is also able to transcend the change or loss of its original function.
We are interested in creating environments that endure, and that inspire flexibility and a sense of freedom.

Fung＋Blatt这样阐述他们的建筑理念：
我们的设计过程开始于一系列相关的同步调查研究。
洞察力＋经验
在我们的作品中，形式的产生来源于对以往经验的总结。
通过提出一些向传统模式挑战的问题，我们重新定义了塑造空间的因素和经验；并且通过这种方式发现一些与物理环境相关的新尺度。
对居民习俗、光照和气候的暂时性变化，材料的触感和音质等问题进行探索，使得它们在整体和内在经验的创造中可以相互衬托。
建筑连接了空间和行为，它也定义了空间和行为的相互转换。只有在不断地转换，复合，重叠过程中才能形成新的体验，熟悉的事物才能以新的形象出现。
场地+结构
我们相信资源的有效利用，材料表现以及结构是组成环境的重要部分。
对位置的静态和动态力量的分析，以及对哪一种力量应当加强或平衡的分析，使得我们可以做出更加有效的决定以确定需要完成什么以及采用何种可行的方法去完成。
这些探索形成了一个结构策略，在这个策略中可以发现一种正式的骨架，并在这个策略基础上来组织空间和整个过程的体验。
变换＋相关性
一个好的设计应当根据时间的变化，考虑到人们如何居住在环境中以及与环境产生何种的联系。虽然我们不知道居住者的过去是怎样的，但是满足他们目前的需要，并且预测他们未来的变化和发展却是我们可以办到的。
一个可靠恰当的环境不但应该非常适合当前的使用需要，而且也应当能够超越它原始功能的变化或消失。
我们的兴趣在于创造持久的环境，并且赋予它一定的灵活性和自由度。

Interview

Project Survey

Fung + Blatt:
In the Architects' House

An interview by Paulette Singley

September 2004

Alice Fung, Michael Blatt, and I are sitting in the dining room of their house in the Mt. Washington neighborhood of northeast Los Angeles, beginning a discussion concerning the guiding principles of Fung + Blatt Architects. The house, completed in 2003, distills, however humbly, many of the themes and theoretical positions that we will be considering in the following discussion. To begin, the expressed structure of steel and concrete, as well as the clarity of interior spaces reflected in exterior volumes, clearly follows the tenets of orthodox modernism. Likewise the apertures, or openings in the walls, frame particular views that both disclose and conceal privileged moments that also might fall within the modernist practice of treating building facades as unique expressions of both interior needs as well as contextual forces. And yet, the house also betrays subtle idiosyncrasies that counter this initial reading and indicate that there is more lurking inside these walls than an idea about the free plan or a reductive kind of functionalism. The house speaks about a working idiom that shifts from conventional to stressed details, from a strict choreography of the body in space to the casual slippage of one function blurring into another -- a floor becomes a seating area, hallways become work areas, and private spaces leak into public spaces. Such shifts betray an interest in creating transcendental spaces where architecture moves us with details and forms as subtle as walls washed with indirect light or materials rendered to retain evidence of their making. Both Michael and Alice are interested in the possibility of architecture performing both overtly and covertly. To enter into the work of Fung + Blatt is to open Duchamp's door into a field of double-functioning elements that perform in multiple capacities, into a spatial milieu of the trace memories of lives lived and architecture's possible mutation over time beyond the abstract realm of pure form, and into the quotidian exigencies of dust, stains, scratches, or scuff marks that perform as significant found objects in the larger material field.

在位于洛杉矶市东北方向的华盛顿山上的艾丽丝·冯和罗斯纳·布拉特家的饭厅里，我与他们开始了关于Fung + Blatt建筑设计原则的对话。这间建成于2003年的住宅旁证了我们在以下的对话中涉及到的许多关于建筑的主题和理论性的观点。作为谈话的开始，钢筋混凝土结构以及反映在外部景观上的内部空间明晰性都明显地遵循了传统现代主义的风格。同样的，像缝隙或者墙上的开口，构成了反映或隐藏专有时刻独特的景观，这也是现代主义对于建筑表达内部需要和周边影响的独特方式。而且，该建筑也显示了精细的特性，该特性计算初次记数并在这些墙里体现了比自由设计思想或实用主义类型还要多的东西。该建筑也表现了从传统的操作到对细节的注重，从空间上的组织到一种功能的不经意的转移——地板成为坐椅，走廊变成工作区，以及私人区渗人公共区。这样的改变也显示了对创造超越性的空间的偏好，在这个空间里，建筑处处显示细节和格式，这与为显示建筑风格而使用间接光线或者材料一样的精细。迈克尔和艾丽丝都对使用那些公开或者隐藏建筑风格的可能性感兴趣。进入Fung + Blatt的作品就像是打开了一扇杜尚式的大门，进入了一个具有多重功效的双功能要素区域，进入了一个追寻旧日回忆的空间环境和超越建筑的抽象的纯形式的突变建筑，也进入灰尘、污点、擦痕等在更大的材料领域起着重要作用的区域。

Fung + Blatt 访谈

地点：Fung + Blatt住宅

被访问者：Fung + Blatt

访问者：Paulette Singley

时间： 2004.9

PS: Why do you want to conduct an interview, a chorus of several voices, rather than an essay that merges your ideas into one voice?

AF: An interview makes sense for us because our work occurs from a process of constant dialogue. Such is the nature of our collaboration. At its worst, as when we first got started, our work bordered on schizophrenia (laugh). After years of honing our process, I would like to think that our work, at its best, strives to produce a tension that holds the architecture in a state of suspended balance. Our work expresses an aesthetic where the components remain distinct and play off of each other. The interview process facilitates a similar tension in the process of writing.

PS：为什么要进行这次多种观点交融的访谈，而不是把你的想法写入一篇论文？

AF：这次访问对我们很有意义，因为我们的工作是来自不断的交流和对话的过程中的，这是我们合作的本质。最糟的情况就是，当我们刚刚开始时，我们的工作就与精神分裂症沾上边了（笑）。在经过多年的努力工作之后，我想说的是我们的工作就是要努力创造一种使建筑物处于一种暂时平衡的紧张状态。我们的工作是要表现一种美学，其中每个组成部分都要保持截然不同并相互衬托。访问也同样是这样一种类似的写作过程。

Designed and built by Michael Blatt for his father, this 350 ft^2 studio sits above an existing two car garage.
The double door opening above the driveway permits access for large materials and lets in western breezes and a view. Primary glazing to the North lets in ample even natural light. Three small wooden shutters to the South, control ventilation and southern light.
The city setback requirements have the second story shifted from the south to rear walls of the existing garage. Lateral forces at the front face of the garage are resisted by a braced strut which is tied to a concrete anchor.
The expressed structure, interpenetration of new and old, play with symmetry, proportioning within a major-minor module explored in this first project are themes that will find their way into later projects to come.

迈克尔·布拉特为他的父亲设计并建造了这个坐落在两个旧车库之上的350 平方英尺的画室。
车道上开着的双重门不仅可以通过大型的物件，同时还能带进微微西风，并欣赏到美丽的风景。北面的落地玻璃窗可以得到充足的自然光。南面3个小木制百叶窗调节着空气的流通和南面的阳光。
城市的退红线要求使建筑的第二层从南面移到了原有车库的后墙位置。车库正面的侧力由连接在混凝土挡土墙上的支柱所支撑。
在第一个项目中出现的这种新、旧建筑相互渗透、对称，主次有比例的结构表达方式在随后的工程中也能看到。

1.dusk view　2. exterior view　3. model　4. west elevation

1. 黄昏视景　2. 建筑外观　3. 模型　4. 西立面

Choreography

PS: Although the idea of spatial choreography is central to your approach to organizing space as a sequence of events that unfold one after another, your work also offers a liberal variety of ways it may be approached or inhabited. My observation is that rather than attempting to script human behavior as a method of spatial reform, apropos of the modernist movement's similar attempt to transform society through space planning, your work allows for multiple dances to occur simultaneously in open-ended spaces that betray little interest in overtly scripting movement.

MB: I do not think that the architect knows what is best for people and I do not subscribe to the modernist manifesto that the purpose of architecture is to move society toward a specific utopian future. Nonetheless, I believe that good design should improve people's lives and I work with the assumption that the human intellect holds unlimited potential for the betterment of the world.

I first learned about architecture in high school, from old copies of Arts and Architecture that I purchased from a used-book store for about 25 cents each. I would buy one issue per week. From these magazines I learned to appreciate modernism as a movement that was capable of offering a comprehensive worldview that integrated all of the arts allied in a common mission. The magazine published stories about the future of mass-produced housing together with art reviews and essays about jazz. It seemed wonderfully naive, but it was also very exciting.

PS：虽然你们把将空间作为一个接一个展开的一系列事件来组织的编排方法作为中心，但你的作品也同样体现了多种灵活的方式。我观察到你们试图演绎人的行为，并将其转化为空间创新的方法，这也是现代主义运动所宣称的通过空间规划来进行社会改造的相似尝试。你们的作品允许多种表现形式同时存在于无限的空间中，这些表现形式显示出明显的动感。

MB：我不认为建筑师已经认识到什么对人们来说是最好的，我也不同意现代主义者所宣称的建筑的目的是把社会改造成一个神奇的未来乌托邦。我认为好的设计应当提高人们的生活，而且我也认为人们拥有无限的使世界变得更好的智力上的潜力。

我在高中就开始从老版本的艺术和建筑杂志中学习建筑， 这些杂志是我以每本大约二十五美分的价格从一个旧书店买来的。我每周买一期。从这些杂志上我学会了把现代主义作为一种运动来理解，该运动反映的是当时全球背景下的各个艺术门类的一个共同目标。该杂志也刊载艺术评论、关于爵士乐的文章以及关于大规模住宅的未来的文章。这听上去很天真，但却让人兴奋。

AF: I do believe that all of the arts and humanistic traditions are intertwined. They learn from each other and pursue a common purpose to understand, interpret and come to terms with the human condition. Innovation, as often as not, comes from the margins where disciplines overlap and through tackling issues from different angles.

The spatial choreography in our work is not a hierarchical one. We create moments and transitions, armatures of relationships that are activated by the participant, that allow for multiple paths of experiences, choices and perspectives from where one can view where one has been.

AF：我相信，所有的艺术和人文传统是互相交织的，它们互相学习，寻求理解和表达，并与人类的环境状态相和谐。创新一般来讲是来自于学科的重叠和从不同角度来解决问题的边缘。

我们作品中的空间编排不是等级划分。我们与参与者共同创造空间的瞬间和转换，这是关于体验、选择和视点在主动与被动间的相互寻求的一个综合通道。

A 750 ft^2 addition to an extensively reconfigured, 1,100 ft^2 1960s residential box.
Strong planar elements reach out to the view and down to the ground to tie the house to its severely stepped lot. Supporting the projecting dining room on a pylon that reaches to the lower yard was a response to the reality that the existing retaining wall could not support the addition. Decking and stairways that fan from the pylon choreograph a punctuated descent into the garden.
750 ft^2 of new living space were added and 1100 ft^2 were extensively reconfigured to create a new organization stressing the public functions of the upper floor while providing privacy for the lower. The living and dining rooms, open to each other, are extended outwards to capture the various views.
A new open-tread flight of stairs descends to a mid-level landing that accesses the main deck, while a series of lower stairs and decks provides a choreographed and punctuated descent into the garden. A study located at the base of the stairs provides a buffer between the private zone and the primary circulation. The entrance to the master bedroom, off of the study, begins a sequential retreat burrowing deep into the hillside.
Interweaving solid and void and existing and new, the house is at once anchored to the ground, while floating above it.

这是一个在20世纪60年代建造的1 100 平方英尺住宅的基础之上增扩750平方英尺的项目。
强有力的平面元素向外伸出，并向下延伸至地面，将房子和地基紧紧地连在一起。为了弥补现有的挡土墙不能够支撑新增建筑物的不足，我们利用了一个伸到下面院子内的塔门来支撑一个外突的餐厅。从塔门处以扇形展开的装饰物和楼梯使得向下通向花园的道路回旋有致。
原建筑新增了750 平方英尺的生活空间，同时还对原来的1 100 平方英尺旧空间进行了全面改造，新的组织结构同时强调了楼上的公共性和楼下的私密性的相互支持。客厅和餐厅对面而开，都向外延伸以获得多样化的风景。
新的露天台阶安置在半高的坡地上，并通向建筑的主平台，同时还有一些低的台阶和平台也错落有致地通向花园。位于台阶底部的书房为私人区和主流通区之间提供了缓冲。书房旁边是主卧的入口并渐渐地伸入到后山。
建筑布局的虚实相间和新旧结合使得这套房子固定在地面的同时又好像飘浮于其上。

1. view of dining room　2~4. interior view　5. view of dining room

1. 餐厅　2~4. 室内　5. 餐厅

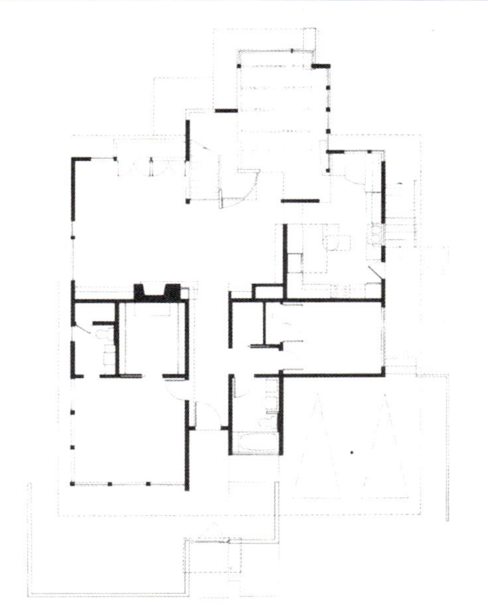

5

6. dining corner
6. 餐厅一角

7. dining looking into living
7. 从餐厅看起居室

8

9

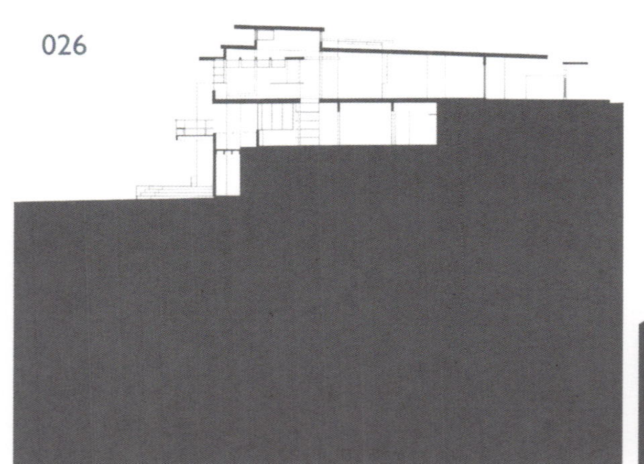

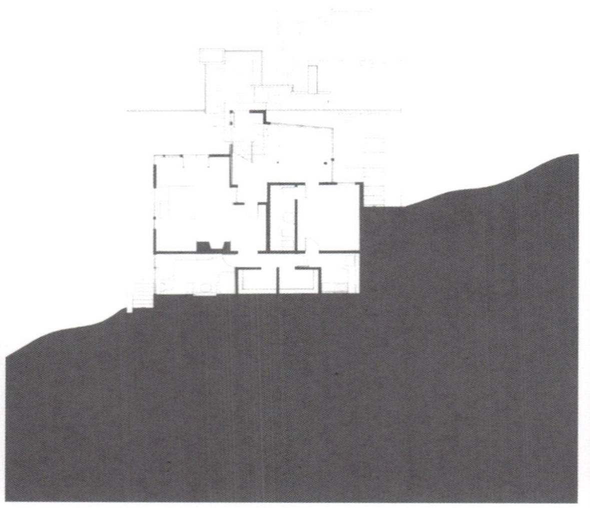

8. view from southwest 9. stair 10. section looking north, lower floor plan and north elevation 11. bathroom 12. view from northwest

8. 从西南方看建筑 9. 楼梯 10. 北侧剖面图，低层平面图和北侧立面图 11. 洗漱间 12. 从西北方看建筑

12

View and Context

PS: According to Hubert Schmalix, a painter whose Mt. Washington house you designed, windows were not a necessary component of the studio. He claims that "I make my own views...the one that is outside is too distracting."[2] Given that Schmalix immigrated to Los Angeles from Vienna, I could not help but to think of the comparison to what Adolf Loos said to Le Corbusier about windows. Beatriz Colomina noticed this passage in Urbanisme of 1925 where Corbusier writes: "Loos told me one day: A cultivated man does not look out of the window; his window is a ground glass; it is there only to let the light in, not to let the gaze pass through."[3] The comparison I see here is that, even though many of your residences are on hillside sites with spectacular views, you are entirely discriminating about where you conceal and reveal the scenery.

MB: In fact, the Schmalix house has windows in every room and the residential section is oriented to the various views. But the design of his studio is all about the light from the clerestory skylights above, which are oriented to true north. We did include one small window, because he smokes and we wanted to provide a moment of relief, a place for him to stand and look out. His paintings are large and he would stand far back to look at them, so we designed the studio to maximize the wall space.

PS：在你们为画家Hubert Schmalix设计的华盛顿山上的住宅中，窗户成了工作室的非必要组成部分。他认为："我有我自己的视角...... 其中一个就是外部环境太令人精神涣散了"。由于Schmalix是从维也纳移民到洛杉矶的，我情不自禁地想到了阿道夫·路斯（Adolf Loos ）与勒·柯布西耶（Le Corbusier）所谈论的关于窗户的观点。Beatriz Colomina在1925年的Urbanisme杂志上注意到这篇文章。在这篇文章中，柯布西耶写到：有一天路斯告诉我：一个有文化的人不会从窗户向外看。他的窗户用的是磨砂玻璃，这样就只能让阳光进入而不能向外观看。我对这种比较的理解是即使许多住宅位于风景壮观的山坡上，但你要能决定什么是应该隐藏的，什么应该是展示的。

MB：事实上，Schmalix建筑的每个房间都有窗户，而且卧室在多个视角都有窗户。但是他的工作室只能从开在正北方的天窗上射人光线。我们设计了一个小窗户，因为他抽烟而且我们也想给他提供放松的时刻，一个对他来说可以站着向外远眺的地方。他的画幅很大，需要站在远处观赏，所以我们通过设计使这个工作室墙壁围合的空间最大化。

We have worked on a lot of hillside homes and find that if we are not careful, a view can be tyrannical; it demands all of the attention for itself. The classic Los Angeles view from the hillside is very powerful but also very monotonous. It offers very little opportunity for interaction with the architecture. Of course one example of a house that uses this context to great effect is Pierre Koenig's Case Study House #22 in the Hollywood Hills. Julius Schulman's famous photograph focuses on the house as seen in context of the view. The grid of the city and the steel frame seem to derive from each other. It's a great architectural moment, but people rarely ever talk about the floor plan of that house.

AF: A dominant view can breed tyranny, in that it tends to celebrate the vantage point of seeing what is below, of visually dominating the landscape instead of being a part of it. The depiction of the view and the house in Schulman's photograph is compelling because the house is represented as a glass cube that projects out into the city grid, establishing a direct orthogonal relationship between the house and the city, the space one occupies and the larger world outside. The house appears to have grown out of the context.

I believe a scale change is required for the view to become relevant. I think about being in New York, with its dizzying array of towers and blocks at the street level suddenly becoming comprehensible from the elevated position of the Empire State Building. That relationship and sequence is key to the relevance of the view. It references you to the larger expanse of the city.

我们设计了很多在山坡上的房子，我们发现如果设计的不够仔细，得到的视角将是很难看的。这需要特别注意。从山坡上看到的洛杉矶美景是有力而单调的。它极少提供建筑的相互衔接的机会。当然，在好莱坞山上的Pierre Koenig的 Case Study House 22号使用的就是这种运用周边环境的效果的案例。Julius Schulman的著名摄影就是聚焦在这种视角下的建筑物。城市的规划和钢筋骨架看上去互相衍生。这是个伟大的建筑瞬间，但是很少有人谈论该房屋的平面设计。

AF：优势的观景方式能衍生统治感，因为这种观景方式倾向于选择向下俯视而取得视觉上的支配地位。Schulman摄影作品中的景色和房屋的描述是引人注目的，因为这个房子是作为一个玻璃立方体呈现的，该立方体设计突出在城市的规划中，在房屋和城市之间以及在房屋所占用的空间和更大的外部世界之间建立了一个直接的正交关系，该房屋显示了其在周边环境中的突出性。

我相信对观景的相关性的要求就是规模上的变化。我想在纽约，从帝国大厦远眺纽约那一排排令人晕眩的大楼和街区，就能瞬间地体会到这一点。这种关系和顺序是景色相关性的关键。你可以从城市的更广阔区域来得到印证。

MB: A view also is a context, one that offers an opportunity to produce an architecture that is greater than itself. Casa Malaparte[4] achieves this brilliantly. It remains a fully contextualized architecture that not only frames views but also has become a part of the view. This is also true of Falling Water. Frank Lloyd Wright put the house where the clients expected to be looking -- the building occupied the view of the waterfall. In the photograph of Case Study House #22, it is not so much the view from the house that impresses, but more so the way in which the house emerges from its context.

AF: Casa Malaparte also questions the role of the windows as they participate in the sequencing of the viewing experience. The heavy frames around the windows operate as quotation marks that bracket the view and transform it into quotations or paintings on the wall. To fully experience the view as a panorama you have to ascend a broad stair that is the building itself. Photographer Karl Lagerfeld, who stayed at the villa, wrote, "Standing on the top of the famous roof-terrace, one feels like a tree looking down at the sea."[5] From that vantage point, the viewer becomes a part of the landscape that unfolds below. This is a great strategy of transforming the view into the context.

MB: In the Dillon Street Residence, as in our own house, apertures/voids are located strategically, to frame the nearby rim of a canyon or to merge with the broad views of the city. Finally, creating internal views from space to space becomes a powerful design strategy. The Loos quotation suggests that it also is possible to discover fundamental truths, as Einstein did, just by thinking inside one's own head and never looking outside.

MB：一个景观就是一个背景，并能显示出比建筑本身更大的东西。Casa Malaparte建筑能达到这点。这是个完全能融入背景的建筑物，它不仅仅是观景器而且还是景观的一部分。这也正是流水别墅的建筑原理，弗兰克·劳埃德·赖特(Frank Lloyd Wright)把房屋建在住户期许的位置，使之成为可以欣赏到瀑布的建筑。在Case Study House 22号的照片里，与其说那所房子令人印象深刻，还不如说使房屋在周边景观中突出的手法更让人难以忘怀。

AF：由于窗户在欣赏景色的顺序上有一定的作用，Casa Malaparte 也质疑窗户的作用。窗户周围的厚重窗套好像是个引号把景观括进来并且将它转变成墙上绘画。为了欣赏全景，你需要站在建筑本身宽阔的楼梯上来体验。当摄影家Karl Lagerfeld站在别墅顶上时，他写道“站在房顶上欣赏景色，就像站在一棵树上俯视海景”。站在这种位置，观赏者就融入了他所俯视景观的一部分。这是把视点转化为周围景色一部分的重要手段。

MB：正如在我们自己房子的空旷处与相邻的溪谷构成一个整体一样，Dillon大街住宅的开口也与城市的景观连成一片。最后，从空间到空间的内部取景成为一个有力的设计策略。根据路斯的建议：正如爱因斯坦所作的一样，只通过自己的头脑而不从外部世界去探索来发现真理也是可能的。

AF: Regarding Loos's statement that the cultivated man only requires a window to let light in, he is describing an introspective, psychological space. Conversely, in the Chinese literati painting tradition, the cultivated man was often depicted as communing with nature. I think that one must strive to create this duality; it can be a powerful thing.
Casa Malaparte achieves this. R.M. Schindler's Kings Road House in Los Angeles also exploited this duality in its embrace of the extremes of what makes shelter – from the ephemeron of an outdoor fire to the permanence of fortified walls. This simultaneous sense of containment and exposure is something we try to bring to our work -- at our own house, in the Maunu Poolhouse, in Kenner Studio, for example.

AF：路斯所认为的有文化的人只需要一个让光线进人的窗户的论述，只是描述了一个内省的，心理的空间。相反，在传统的中国文人画里，有文化的人常常是融人自然的。我想一个人必须努力去糅合这种二元性，它将是最厉害的。
Casa Malaparte建筑做到了这一点。位于洛杉矶的R.M. Schindler国王路住宅也同样探索了在这种二元性里如何做到最极端的密闭——从室外大火的短暂性出现到防御墙的永久性。这种密闭和开放同时发生的感觉就是我们在自己的作品中极力营造的东西——你可以在Fung+Blatt住宅、在Maunu Pool住宅和Kenner工作室等项目中感觉到它的存在。

The Schmalix residence was designed to house the studio and family of Hubert Schmalix, a renowned Austrian artist and founding member of Austria's "Wild" painters. The residence is located on Mount Washington, a century-old bohemian enclave that rises at the confluence of the Los Angeles River and the Arroyo Seco. The house nestles into a hilly half-acre of native wilderness, commanding broad, open views of the Arroyo Seco and Downtown Los Angeles.

The first steel-frame house designed by our studio for Mount Washington, the Schmalix residence marks a return to innovative architecture in a neighborhood that boasts houses by Richard Neutra, Rudolph Schindler, Gregory Ain and Harwell Harris, as well as many significant Craftsman homes from the turn of the century.

From his former home in Mount Washington, Hubert Schmalix created a series of paintings of the sprawling basin floor below. The ubiquitous pattern of roof-tops, bisected by asphalt strips, power poles and cypress trees is isolated and abstracted into a formal dialogue that registers, at once, familiar and alien. The Schmalix Residence abstracts its materials and forms from the vernacular housing and industrial buildings of the basin below. Its details, however, express a structural backbone that finds commonality in its modern and craftsman local precedents.

Schmalix 住宅是为奥地利著名艺术家，“旷野”派的创始人之一Hubert Schmalix设计的工作室和家居。这套住宅坐落于有百年历史的波希米亚人的领地——华盛顿山脉，位于洛杉矶河和Seco河床的交汇处。这套住宅占据了半英亩原野，为群山所环绕，视野开阔，居高临下可以看到seco河床和洛杉矶市区的景象。

位于华盛顿山脉上的Schmalix住宅是我们设计的第一栋意味着向创新建筑回归的钢结构房屋，这栋建筑可以与19世纪末20世纪初的许多知名艺术家的房子，如邻近的Richard Neutra、 Rudolph Schindler、 Gregory Ain and Harwell Harris引以自豪的房子相媲美。

在Schmalix 的华盛顿山旧居室里，他创作了许多关于盆地农庄的油画。那些随处可见的被沥青带、电线杆以及柏树林一分为二的屋顶。这些抽象的分割让人好像进入了一种形式的对话，感到似曾相识。Schmalix住宅从盆地民居和工业建筑中提炼出材料和样式。然而，其细节表现了结构的精髓，展现了现代和地方工艺相结合的特点。

Several themes in Hubert Schmalix's paintings interest us in their application to Architecture. Among them are the interplay of depth and surface and the regenerative use of forms. This thematic substrate led us to explore transparency and opacity, transiency and permanence of materials; as well as the interplay of organic movement against a formal armature of modular components.

Built to the rigorous standards of Los Angeles' new seismic and hillside requirements, the Schmalix residence employs common industrial and renewable materials.

Light gauge steel framing in the building allows floor spans and wall openings not usually achieved in domestic construction. The house's assembly of steel, concrete and galvanized sheet metal is expressed inside and out. Curtain walls of glass and bearing walls sheathed with polycarbonate fulfill the need for good, modulated light and the desire for spacial extension into the landscape; while massive, bare concrete foundation walls emphasize the house's connection to the land. Complexities that occur where forms and materials overlap lead to exploration of relative transiency and permanence.

From the street, the house is a linear composition of three structures. To the north, a linear mass with a skewed saw tooth roof houses the studio. To the south, the entrance into the living quarters is carved out of a quonset-like structure which shelters a study and extends into a slope over a double height living space. These two volumes, bridged by the carport roof, crank against one another as the hillside wraps around, preserving view from the street through to the canyon below.

While the Living Room mass extends out towards the view, the Kitchen and Dining/Family Area are tucked under the Study and the Carport respectively against a board formed concrete retaining wall that runs the length of the building. The expanse of the concrete is eased by a curved wall, swelling from the Kitchen through the Dining Area into the hallway, that conceals the service functions of the house.

Alternating bedrooms and baths break loose from the hallway into a dynamic infill zone that slides free under the rectilinear mass of the studio. Each Bedroom orients outward to a differ.

Hubert Schmalix的绘画作品中的一些主题之所以吸引我们是因为其在建筑上的应用，包括对深度和表面的相互作用以及对形式的创新运用。这种主题基础引导我们进入对材料的透明和不透明及材料的暂时性和永久性的探讨，同时也引导我们对于有机运动和模式元件之间的相互作用进行探讨。

Schmalix住宅采用了普通的工业材料和可再生材料，严格按照洛杉矶的新防震标准和山地建筑的要求建造。

轻型钢质框架在建筑中的使用使得楼层的跨度和墙体的开放度超越了一般意义上的建筑。房子的内外都可以见到钢材、混凝土和镀锌薄钢板。玻璃幕墙以及聚碳酸酯质的移动墙可以满足对于采光的调整需求以及空间上向外延伸的需要。同时，大量裸露的混凝土房基加强了房子和大地的连接。材料和形式重叠处的复杂性开启了对暂时和永久的相关性的探索。

从街道上看，这栋房子是三个结构体的线性组合。从北面看，一排倾斜的锯齿形屋顶覆盖着工作室。从南面看，起居室的人口镶嵌在一个活动的建筑结构里，它将书房掩盖其中，并且延伸到有起居室两倍高的小斜坡上。这两个建筑体通过底下车库的屋顶连接在一起，形成迂回的势态像周围小山一样相互环绕，将从街道到下游峡谷的景象尽收眼底。

在起居室向外延伸的同时，厨房和餐厅隐藏到书房和车库的下面，分别倚靠着环绕整栋建筑的混凝土挡土墙。宽阔的混凝土墙体因墙面的弯曲而显得灵活，这堵弯曲的墙面经过厨房、餐厅和走廊将房子的服务性功能掩于其中。

卧室和洗浴室的相互交错打破了从走廊到动态填充区的松散，这个动态填充区像是从画室轻松地直线滑落下来。每间卧室都面向室外不同的方向。

1. street elevation　2. studio　3. upper floor plan

1. 邻街立面　2. 画室　3. 上层平面图

2

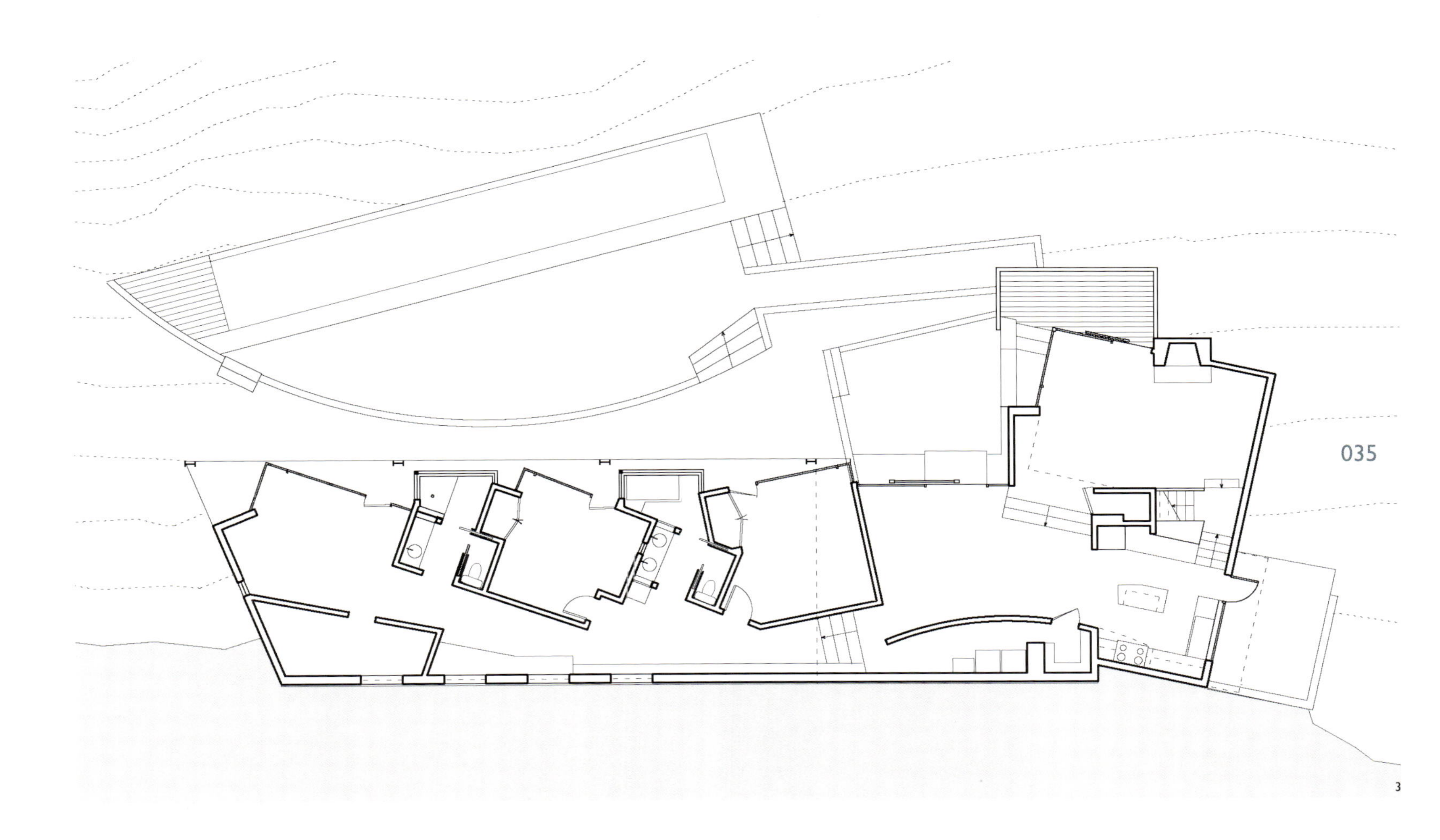

3

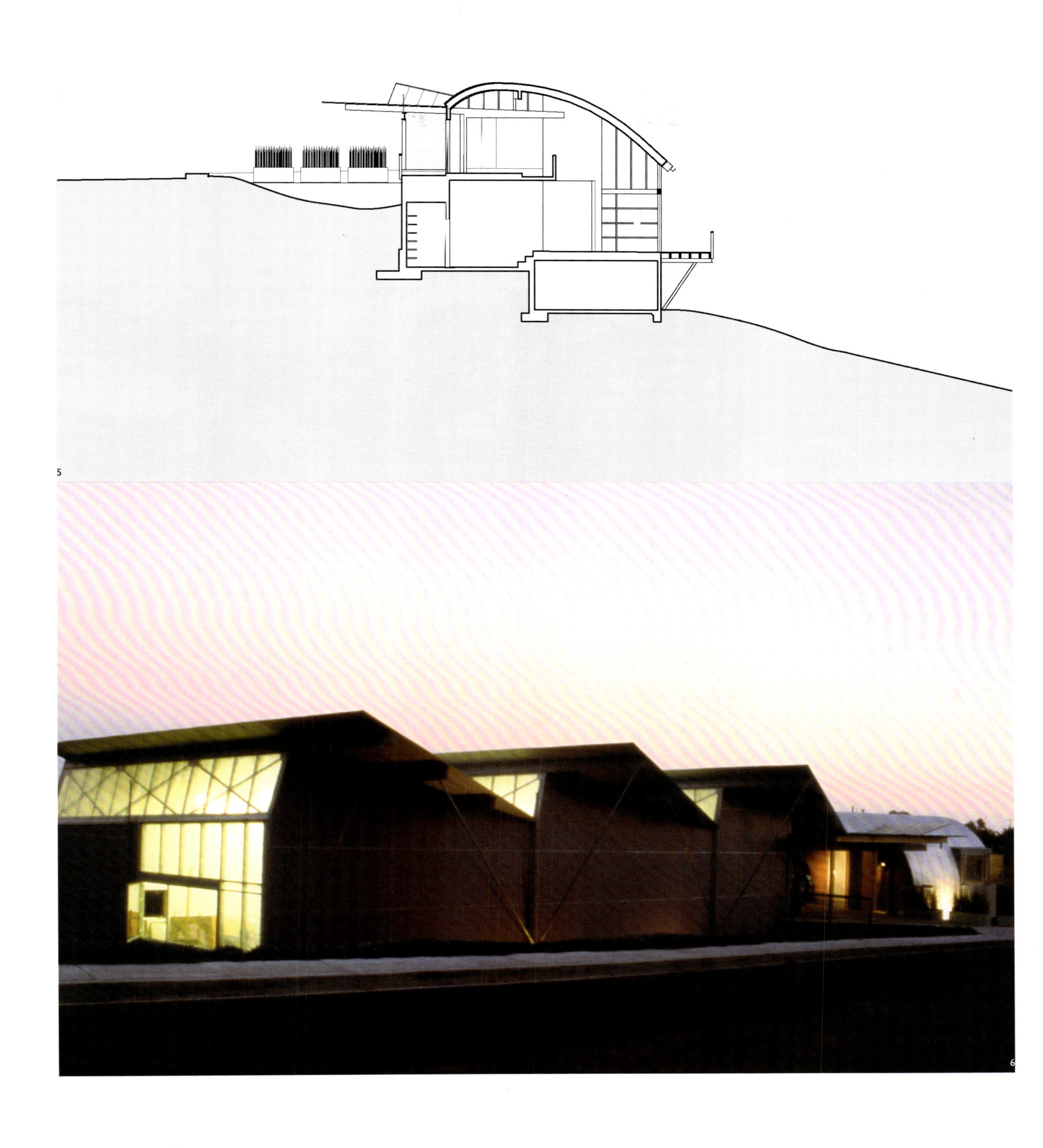

4. exterior view　5. section through living　6. house front dusk view

4. 建筑外观　5. 起居区剖面图　6. 黄昏中的建筑前景

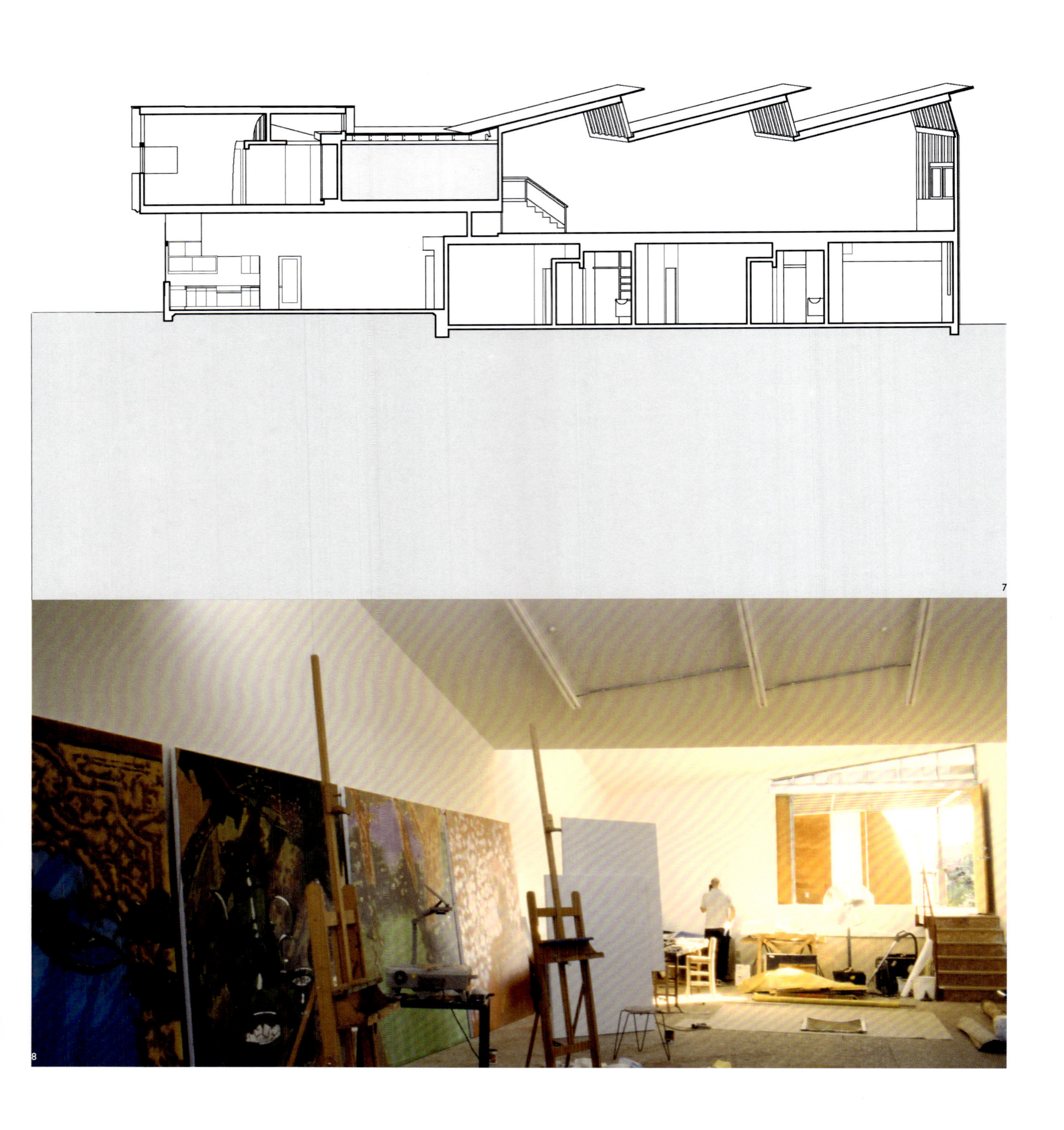

7.longitudinal section　8. studio　9. house front　10. living　11. bath room

7. 纵剖面　8. 画室　9. 建筑前庭　10. 起居室　11. 洗漱间

9

10

11

12. master bed room　13. kitchen

12. 主卧　13. 厨房

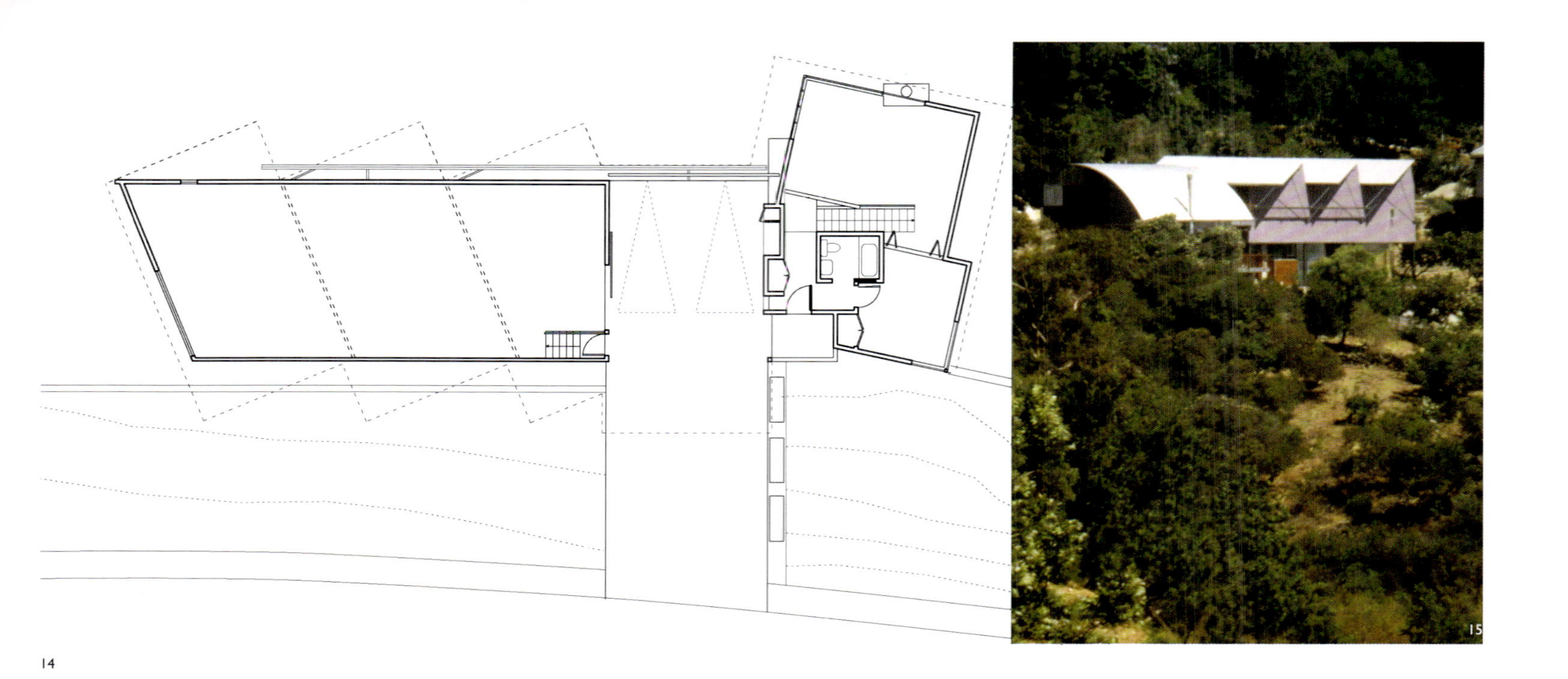

14

15

16

17

18

14. lower floor plan 15. hill side view 16~18. views of hallway 19. rear view

14. 低层平面图 15. 丘顶一侧 16~18. 室内走廊 19. 建筑后部

19

Duchamp's Door

PS: A certain aspect of your work, the elements that serve a double duty such as the dining room floor that extends out from the concrete slab that turns into the top of a wood seat which also serves as a stair that also opens into a storage cabinet, makes me think of the door in Marcel Duchamp's Paris apartment that served simultaneously as both the bathroom and the front door. When the bathroom door was closed the front door was open and vice versa. When everyday objects perform a double duty, even if it is just for practical purposes, the effect is that they are perceived in new ways heretofore unknown. The adjacency of functional doubling creates formal de-familiarization as does the displacement of objects from their seemingly natural contexts. To what extent is this design strategy a result of intellectual conceit or simple pragmatic restraints?

MB: Apparently Duchamp wanted to contradict a French proverb that a door is either opened or closed. I have sat down to design a Duchamp door, and it is very tricky to do. You can do it with a pivot hinge. It has made me think a lot about what exactly a door is. I have been trying to find a place to do his door since graduate school, but I have not yet encountered a practical reason for it. Alice and I share a practical and a formal sensibility yet we are attracted to ambiguity or perceptual perversion. Something that is purely intellectual is not entirely satisfying because it is all about itself. We create these elements that perform double duty with practical justifications, knowing full well that we also are doing this also because we like the ambiguity. Of course, this kind of work is appealing on a lot of different levels. It is very satisfying to solve a problem in a clever way, but with these kinds of solutions, we can also transform the way we relate to space and challenge convention.

PS: Your double-function elements have the potential to create a disturbing or strange encounter with the object that appears to have roots in the Surrealist movement.

AF: While traveling in Europe this summer, we saw a Magritte painting titled "The Rape" where, through figure/ground reversal, the victim's and rapist's shadows merged into one figure. Chilling. Gerhard Richter did some interesting studies on perception. We're seeing rather surreal details in the Jugendstil architecture around town -- playing with perception and triggering multiple readings. We're definitely into that with our work. Circularity, shifting points of reference, and reversal of interior and exterior really extend to how we see our world.

PS：元素间的双重用途设计是你们作品中的重要特色，如沿着餐厅的地板向外延伸的水泥板转换为既是楼梯又是通向储藏室的坐椅，这使我想到同时作为浴室和前门的马塞尔·杜尚的巴黎公寓的大门。当浴室的门关着时前门就打开了，相反亦然。当日常物品有着双重功能时，即使只是为了实际的使用目的，也会产生以前所未认知的视觉效果。由于它们现在的功能是从原有的功能转换而来的，这种建立在物体表面属性上的双重功能的置换产生了一种新鲜的感觉。这种设计策略到底是创新的想法的结果还是只是简单的实用主义的约制呢?

MB：显然，杜尚想去反驳一个法国谚语，那就是："一个门或者是开着的或者是关闭的"。我已经设计了一个杜尚门，它非常的巧妙， 你可以使用一个铰链来完成它。这使我对门是什么这个问题有了更多的思考。自研究院毕业后，我一直努力寻找机会来建造这样的门，但是我一直没遇到一个来实践它的理由。艾丽丝和我都拥有实践和对形式的洞察力，但我们还是被模糊和感性的滥用所吸引。一些看似智慧的事情不能让人完全满意还是因为其本身的原因.我们创作了这些能在实际中有效地实施的双重功能元素，因为我们完全认识到我们所作的，也因为我们喜欢这种模糊性。当然，这种工作可以在不同程度上显示出来。它是解决问题的有效手段，这些解决方案也是我们改变空间和挑战传统的方式。

PS：你们的双重功能元素可能会创造出一些令人惊异的或是奇特的东西，这些东西与超现实主义的创作不谋而合。

AF：我们今年夏天在欧洲旅行时，看到了一件名为"强奸"的马格里特的画作。其画作通过形状和场所的颠倒，受害人和强奸者的阴影混合在一个图案中。很冷漠。Gerhard Richter对感知做了一些有趣的研究。我们在新艺术建筑的城镇周围里看到了超现实主义的细节——运用感知去触发多重理解。这些也明确地存在于我们的作品中。循环、坐标的变换、内部和外部的颠倒表达了我们对世界的看法。

MB: My primary influences are the principles of social utopianism and the aesthetic constructs of surrealism and psychedelia. I mention these specifically because of the way they play with perception and emotion. I have been influenced by de Chirico and Magritte and am impressed with the multi-sensory aspects of psychedelic art. I operate with an underlying belief that the avant-garde movements of all disciplines must inherently be aligned. By producing effects which are familiar yet alien, disturbing in their implications while reassuring in their verisimilitude, they allow for the reexamination of the world free from the typological constructs which limit our thinking.

AF: We find interstitial spaces, such as the residual plazas and paths left from the figure of the buildings, to be some of the most interesting places of an urbanscape. In a home, spaces where areas overlap and where one function spills into another create ambiguous boundaries that allow for the greatest freedom to reinvent.

The Schmalix house has a long path/hallway down the bedroom wing. We believe hallways, in and of themselves, are a waste of space. So playing off a long concrete retaining wall, we activated the opposing wall. The hall swells at several points: at the foyer, at the entrance to the children's bedrooms and a bath, at the other end for a computer area, and before the master bedroom. As in our own house, we located the sinks in areas directly open to the circulation space, compartmentalizing only those functional areas that require privacy -- the toilet and bath area. On one level, this decision stems from the desire to isolate materials and objects, the components from each other, as well as an attempt to open up the space and to emphasis the purer diagram of the building. At still another level, it celebrates the ritual performance of washing one's hands. Thus it is important for such "perversions" to work on several different levels, that they serve an integrated function as well, such that you don't notice them per se, but that they nonetheless alter perception.

MB：对我影响最大的是社会乌托邦和超现实主义与迷幻混合的建筑美学。我特意提到这些是因为他们处理感觉和感情的方法。我受到基里科和马格里特的影响并被迷幻艺术的多个感官外表所吸引。我潜意识认识到所有学科的前卫艺术有其共同之处。通过产生一些相似而不是相异的效果，并在保持其逼真性的同时混淆他们所暗示的，他们让我们从限制思想的传统建筑模式中解放出来得以重新审视世界。

AF：我们发现空隙空间，如大厦的剩余空间和建筑所产生的路径成为城市景色中最有趣的地方。在房屋里，重叠的和同时充当双重功能的空间区域是能产生最大创新自由度的模糊边界。

Schmalix住宅沿着卧室边有一个长长的走廊空间。我们认为走廊是空间的浪费，所以设计了一个长的混凝土防护墙，这使得相对的卧室墙被我们激活了。建筑在好几个方面变得很开阔：在门厅里、在孩子卧室和浴室的人口处、在计算机房的尽头、在主卧室的前面。正如我们自己的房屋一样，我们直接在开放的循环空间中安装洗漱槽来与那些需要私密化的功能性区域区别开来——卫生间和浴室。在一个层面上，这种设计遏制了材料和物体相隔离的欲望，它们相互作用的同时也尝试了对空间的开放以及强调了建筑的纯粹性。在另一个方面，它也有洗手的功能。你可能没有注意到它们这些多重功能，但是它们却能改变感知，这对于从不同的方面来理解我们的工作是很重要的。

a deck

A two-story deck created to perch to the northeast of an existing stilt house. A boat in the treetops, it serves as an outdoor dining room.

a bath

Redwood evokes the warmth of the sauna, while glass mosaic tiles reflect the light that showers from a skylight above the bath basin. A casement window provides a horizontal view out to the garden. Honed slate and maple further complete the sensory palette for this master bath.

a pool house

Constructed from recycled materials (glass, hardboard, corrugated steel, and redwood), this pool house addition onto a former stable provides bath facilities for pool and guests, while exploring the notion of tenuous enclosure.

平台

一个两层的平台被设置在原有吊脚房的东北方。安置在树梢处的船形屋可以当作户外餐厅。

浴室

红木唤起了人们洗桑拿时的温暖，同时琉璃瓦和陶瓷锦砖反射了从位于浴盆之上的天窗洒下的阳光。透过垂直的铰链窗可以一览花园的景象。光滑的石板和枫木更进一步地增强了主浴室的色彩感。

泳房

这个在旧的马廊上由可再生材料（玻璃，硬木板，波纹钢和红木）加建的泳房为客人提供了洗浴设施的同时也探索了低密度天井的观念。

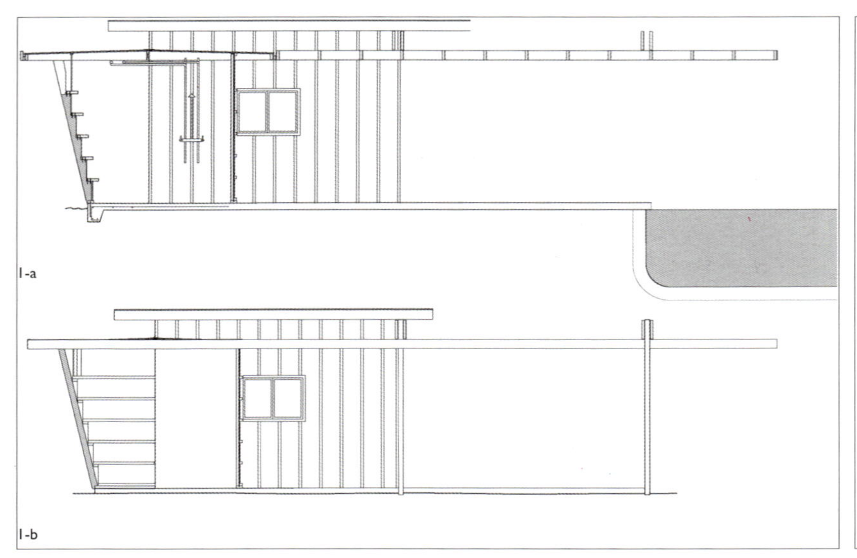

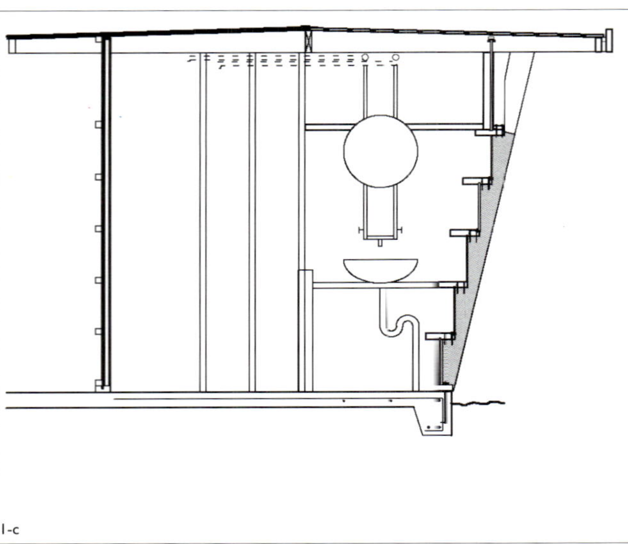

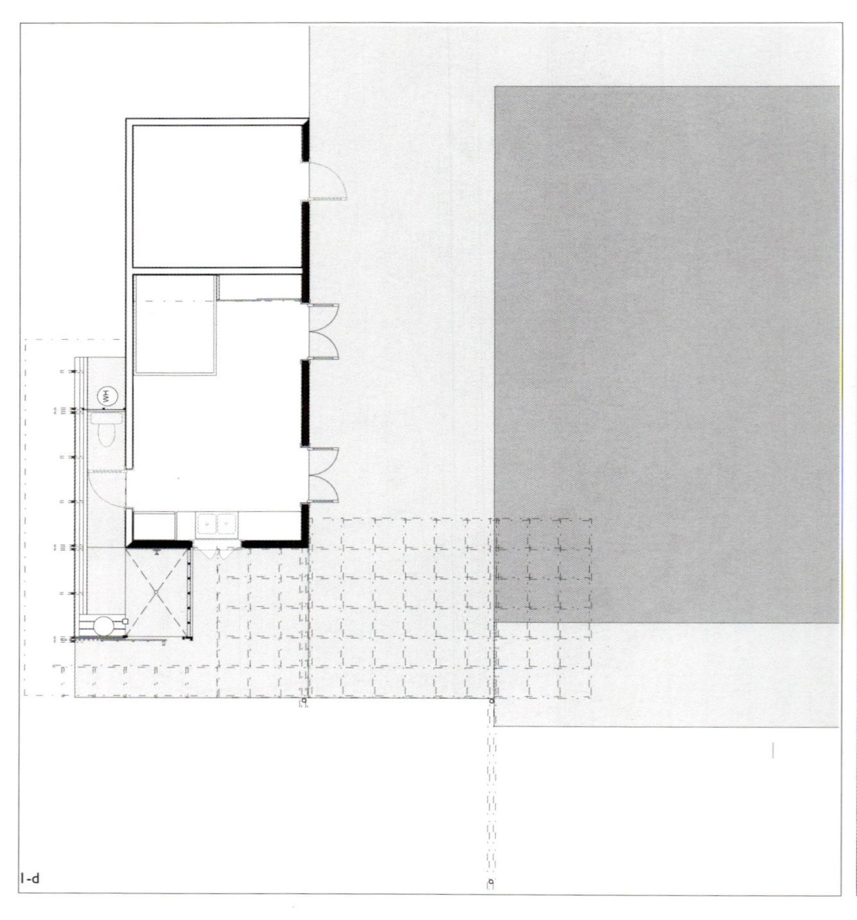

1-a. section　1-b. elevation　1-c.section detail　1~d. plan　2. lantern　3. view from neighbor's

1-a. 剖面　1-b. 立面　1-c. 剖面细部　1~d. 总平面 2. 吊灯　3. 从邻侧看建筑

3

4

4. window　5. shower with skylight　6. bath
4. 窗户　5. 带天窗的浴室　6. 洗漱间

7. eggcrate trellis over portion of pool
7. 游泳池的吊顶

8. pool house bath
8. 游泳池的洗漱间

9. guest room

9. 客房

10. view of shower

10. 浴室

In remodeling this 1950s one-story stilt house, we analyzed an old pattern of use and re-articulated the zones for new and existing uses within an open plan.

A hierarchy of experiences, beginning with a dropped ceiling at the Entry, is introduced. Notched to address three 8 foot high doors, the ceiling becomes a light shelf that leads into the Living area and the view. A soffit traces an existing circulation path along an edge of the Dining area and into a sky-lit hall that terminates at a Bedroom and Bathroom.

A low partition that momentarily redirects a path through an open Kitchen demarcates a family work station. A built-in couch anchors the far end of the Living area.

Skylights, cabinetry and color punctuate and articulate the interweaving the new spaces.

在重新改造这栋建于20世纪50年代的一层的高脚房子的过程中，我们分析了房子的旧的用途，并重新做了一个开放性的区域规划把该建筑的新用途和老用途衔接起来。

为了使这套居室有层次感，我们在进门处设计了一个引人式的吊顶，并嵌装了三扇8英尺高的门，灯光从此处的顶棚照射下来并可以照见起居室。一个拱顶沿着餐厅区边缘原有的环形小道通向了卧室和洗浴室尽头的有着天窗的大厅。

一个低矮的隔断改变了经过开放式厨房的小路的方向，将其与家庭的工作区分隔开来。在起居室的最末端安装了--个固定的睡椅。

天窗，家具和鲜明的色彩赋予了新空间极强的层次感。

1

2

3

1. family work station　2. entry detail　3. entry foyer　4. dining and entry viewed from kitchen (family work station to the right)

1. 家庭工作室　2. 人口细部　3. 人口休息厅　4. 从厨房看餐厅和人口

Trace Memory

PS: Your buildings seem to recall a memory of the site, as it existed before you began construction. What are your thoughts about the relationship between architecture and cultural memory?

MB: The first project I designed and built on my own, a studio addition, was for my parents. There was a lot of personal memory in this project. But even if we were to have torn the entire house down there would still be traces left. You would still have the edges of the house that would indicate where the doors were located, the remnants of the garden, the path of the driveway, and so forth.

When we design our intention is to bend what already exists into something new rather than coming up with an entirely new form that erases the memory of the site.

Our earliest projects were remodels, but when we eventually began completing ground-up work we realized that all projects are remodels. There is no such thing as the blank piece of ground. We approached the remodels with the idea of knitting the new into the old. We produced additions where you could clearly identify that there was new and there was old, but you couldn't tell exactly where one met the other.

PS: Yes, I read that you did this by preserving the opening at the carport of the Schmalix House and retaining a memory of the lost view on the tile frieze on the public storage project in Glassell Park.

PS：你的作品看上去能引起对一个地方的回忆，这种回忆是在你开始建造之前就存在了。你如何看待这种建筑和文化记忆之间的关系？

MB：我独自完成的第一个项目是为我父母建造的工作室。这个项目带有许多个人的记忆，即使我们摧毁掉整个的房屋，记忆仍然留存。你仍可以从房屋的边缘寻找到原来的门是在哪里，花园的残余，车道的路径等等。

我们的设计意图是在已经存在的东西上创造新的，但也不是完全创作一个抹掉过去所有记忆的全新形式。

我们最早期的作品都是改造项目。但是当我们最终开始建造全新的建筑时，我们认识到所有的建筑其实都是改造建筑。这里并不存在像地面一样空白的事物。我们按照把新的东西注入旧的东西的思想来改造。我们制造那些附属物能使你清楚地辨别新旧，但是你不能辨别其确切的连接处。

PS：当然，我知道你是通过保留Schmalix建筑的车库开口和保留Glassel公园里的公共仓储项目中对瓦砾的回忆来做到这些的。

MB: Yes, the carport. That was a memory from when I delivered pizza as a kid. I remember on Mulholland Drive being able to see through pieces of some of the houses that otherwise blocked the view.
AF: At the Schmalix Residence, it is the memory of the cul de sac before the house, where people would park and look out at the view. We were intrigued by the possibility of creating a lover's lane right in the carport where the clients conceivably could sit in the car, drink beer, and look out at the view – a very LA thing to do. It was, as well, an idea about offering the view back to people passing by. We acknowledge the memory, not in a sentimental or representational way, but by way of some form of enactment of the memory.
MB: The public storage project is much more literal than the Schmalix carport. It represents the view that existed on the site before the building was built, although heavily abstracted. The lower portion of the facade is painted to represent the hills, then there is the mosaic that represents a line of trees along the ridge, and there was going to be just the raw concrete above this. But the neighbors wanted it to be painted blue to blend with the sky. But the sky is no one particular shade of blue, so we changed the color at every concrete panel so that at any given point of the day one of the blues might match.
AF: Kim Abeles is an artist who works with smog. She constructs cards with perforations to hold up to the sky into order to measure its relative blueness. This piece on the wall of our house is one of her smog collectors. She infilled the Lascaux bulls with smoke from the 1992 Los Angeles Uprising that had adhered to a sensitive coating on the paper.
But back to the question of memory, we have this brick pillow[6] made by a Turkish-American artist named Ali Acerol He builds the armature of these pieces out of brick and shapes them into this soft kneaded kind of look. They are compelling because they embody the memory of forces acted upon them, be it a pneumatic grinder or the rolling of waves against the floor of the ocean. That ambiguity is key. The pattern of brick that remains just makes you aware of yet an earlier memory, that its origins were once deliberate, perhaps utilitarian. Such is the beauty of ruins. We always hope that if our structures don't survive centuries, they will at least make beautiful ruins.

MB：是的，是这个车库。这使我想起小时候送外卖比萨时在Mulholland大街看到的一些层叠的建筑景观。
AF：在Schmalix住宅中，建筑的前景有着孤岛的记忆。人们在这里驻足并观看景色。我们在车库右边设计了一个情人小径，在这个地方，客户能坐在车里，喝着啤酒，看着景色——这种最洛杉矶式的生活享受。它同时也是个供路人欣赏的景观。我们对记忆的认知，不是用一种伤感的或者具象的方法，而是借用记忆发生的几种形式来认识它。
MB：公共仓储项目是比Schmalix住宅更实际的项目。虽然非常抽象，但它代表了存在于建筑物建造之前的这个地方的景观。建筑物正面的较低的部分被画以丘陵，然后用镶嵌图案表现了沿着山脉的一行树，在这些上面是未经处理的混凝土。但是街邻们想让它涂成天空似的蓝色。但是天空不仅仅是一个简单的蓝色，所以我们改变了每一个混凝土面板上的颜色，使其在一天中的任何一个时间能和其中的一个蓝色相匹配。
AF：Kim Abele是个创作烟雾作品的艺术家。她创造了带孔的叠片以呈现天空并表现它的相对蓝色。我们的房屋墙上的叠片就是她其中的一个作品。她从1992年洛杉矶骚乱事件开始创作依附于纸上的敏感烟雾产生的场景雕塑。
再回到记忆这个问题上，我们有一件土耳其美国混血画家Ali Acerol制作的砖状枕头，他制作了这些非砖块建造的甲胄并把它们纳人软性的景观当中。这些引人注目的外壳是记忆力量的外化，它成为一个风动磨碾或者是海平面上翻滚的波浪。这种模糊性是关键，只是你意识到的早期记忆中的砖块模式，它的起源是深思熟虑的，也可能是功利主义的。这是一种废墟的美丽。我们一直希望如果我们的建筑物不能存在几百年的话，它们至少能是美丽的废墟。

A community spearheaded remediation of a new, vastly oversized self-storage facility at a major vehicular gateway into the community.
Charged to mitigate the apparent mass of the structure, from the street and from the hills above, 90,000 ft^2 of painted mural, tile mosaic, and sculptural steel trellis work were deployed. The mural is an abstraction of the hills and the changing sky. Playing off the panelized tilt-up structure, it evokes a view that was obliterated by the new facility.
The result visually fragments the mass of the building, bringing it down to scale. It effectively erased an eyesore and returned, albeit symbolically, a view that was taken away.
The project now stands as a welcoming gateway into the community, as well as a pleasant reminder of what neighborhood activism can achieve.

此项目是将一个通向社区主要车干道上的私人储藏室设施扩建改造成为一个新的，更大的公共建筑。
为了减轻来自于街道和上面山上对于该建筑的外观压力，我们装配了90 000平方英尺的壁画，镶嵌了瓷砖和雕花的钢格架。壁画是对小山和不断变化的天空的抽象表达。将其装饰成向上倾斜的嵌入式结构，这些壁画带来了一种好像被新的设施所湮没的景象。
这样产生了将整个房屋分解成碎片的视觉效果，好像将房屋的规模变小了。这栏很有效地除去了以往我们不愿看到的视觉垃圾，虽然这是象征性的。
现在，这个建筑伫立在此就像是该社区的迎宾楼，同时它还提示人们如何在社区活动里取得更好的成就。

1.storage crates 2.detail 3. details 4. panorama 5~6. details

1. 储藏条板箱 2. 细部 3. 细部 4. 全景 5~6. 细部

GLASSELL PARK
Public Storage

The 800 ft^2 addition was added to a mid-20th century post-and-beam house. The studio is a 20' by 40' space sited behind the low plane of a parking trellis. A board and batten wall that extends perpendicularly from the entry of the house becomes a concealed entrance into the studio. A north-facing clerestory that rises to nearly twice the height of the house runs the length of the studio. Its apparent height is scaled down by a low eave element and a garden court that separate it from the house.

这是对一个建于20世纪中期柱梁结构住宅的800平方英尺扩建项目。该工作室位于一个停车场墙边低地后面的20英尺×40英尺大小的区域上。一堵从房屋入口处垂直伸出来的木制墙为工作室制造了一个隐蔽入口。

一扇面朝北，接近房屋两倍高的天窗将工作室环绕。一个较低的屋檐结构和一个与房屋分开的庭院降低了工作室的外观高度。

1. carport / studio entry
1. 工作室入口

2. view of studio from street
2. 从大街看工作室

3. detail　4. studio interior
3. 细部　4. 工作室室内

5. view of studio from main house

5. 从主要建筑看工作室

Anderson Residence

Located in a small wooded valley, the Anderson house began in the early 1900s as a one-room cottage, and through the decades, grew into a modest agglomeration of additions and remodels.
In designing our generation's iteration of the house, we found it important to preserve its history, as we reorganized its internal relationships and provided new opportunities for relating to the landscape.
The new Master Bedroom Suite and Guest Room, accommodated in a new second story, rises as a series of roofs above the low-slung body of the original house. The addition is set off from the lower floor by a wrap-around skylight and clerestory that articulates a spiraling path, unfolding as it captures light and framed views of the surrounding hills; and culminating in the master bedroom which opens to a long view across the valley.

坐落于一个树木繁茂的小山谷中的Anderson住宅在20世纪初是以单间别墅开始建造的。然而，几十年过去了，这栋房子已经成为了几经增扩和改造的综合体。
在设计这座经历了不同时代反复修改的房屋的过程中，我们不仅将其内部结构重新组织，而且还认识到保存它的历史的重要性，这是一个为它和风景区之间的连接所提供的新的契机。
新的主卧套间和客房设计在新增的第二层，它们就像是从原来房屋主体上抬升出来的屋顶。新增物被下一层的环形的天窗和通风窗包围着，这些天窗和通风窗形成了一个螺旋通道，并向外伸展，好像它要将阳光和周围小山构成的景象全部捕捉进来；绝妙之处是在主卧室里视野可以透过溪谷达到远方。

1. master bedroom 2. transition from old house to new family room 3. family room

1. 主卧室 2. 新建的家庭室与老房子连接处 3. 家庭室

4. view into family room 5. new stair and continuous skylight above articulating the new from the old 6.downstairs bedroom with articulated beam supporting 2nd floor

4. 看家庭室 5. 新楼梯和上部的连续天窗明确界定了新旧建筑空间 6. 下层卧室用连接梁支撑二层的楼板

POWER. FREEDOM.

7

7.movable screen at new guest room　8. master bath

7. 客房的可移动门　8. 主卫生间

Tectonics

PS: Another way to look at your work is through the crystal clear way materials come together and the innovative, indeed daring, way you express structure. How, then, do tectonics discipline your work?
MB: In designing spaces, the primary palette I deal with is the physical structure. This is not because this is the primary subject, but it is in order to achieve grounding in reality and persuasiveness. Materials remain neutral until you transform them into architecture. Up utill now I have engineered almost all of our buildings. This process allows us to do the "what you see is what you get" kind of thing.
My working on the engineering allows us to do what we want to do on very modest budgets. We also attempt to complete work that not only expresses itself exactly but also that had not been done before. I am always trying to find a new way to make things stand up, of solving something in a unique way. It's a bit of a paradox really: we try to express what is actually going on structurally, while at the same time making the whole thing seem rather improbable. This interest emerged as a significant dimension to our work. This interest also refers to what we discussed earlier, our desire to push people to re-examine their assumptions about the world around them, of encouraging people to stop and think.
AF: We try to look at detailing by gleaning possibilities from unexpected resources, by applying them in unconventional ways that elevate their best attributes. In order to complete the Schmalix Residence and our own house we used curved corrugated metal roofs fabricated at a local prefabricated water tower and tank company. At the Yale-Maclean Residence, we adapted HUD's "Value Engineered Framing" concept that is associated with very inexpensive public housing projects.

PS：从你们的作品里可以看到一种大胆创新的材料组合和结构表达方式。那么筑造学在你的作品里是怎样得到体现的呢?
MB：对于空间设计，我首先是要处理物理结构。这不仅仅是建造的关键问题，而且是达到现实性与可信性的首要条件。材料在被应用到建筑中之前是中性的。到目前为止我担当了差不多我们所有作品的工程设计。这个过程使我们能明确地看到事物的问题所在。
我在工程方面的经验使我们能在非常紧张的预算之内做到我们想做的。我们也尝试完成那些不仅是自身真实反映而且是以前也没被做过的工作。我一直努力试图寻找一种新的途径能让事情经得起考验，并且以独特的方法来解决事情。事实上这有点自相矛盾：一方面我们努力表达结构的真实性，另一方面又要让这看起来不是那么像。这是我们的建筑作品表现出来的一个重要的兴趣点。它同样也涉及到我们刚才讨论过的问题，我们试图推动人们重新审视关于他们周边环境的表现并鼓励他们停下来思考这些问题。
AF：我们通过从意想不到的资源中来寻找细节表达的可能性，以及通过非传统的应用方法来提升它们的最佳品质。为了完成Schmalix住宅和我们自己的房屋，我们使用了一个由当地水塔水箱预制公司制作的弧状金属屋顶。在Yale-Maclean住宅中，我们运用了美国住房和城市发展部有关公共经济房建筑的“价值工程架构”理念。

We are interested in the expression of materials as a kind of chronicle of the building's making. At our own house, the concrete foundation walls, the heavy steel structure, the light gauge steel skeleton, the flesh of the plastered walls, the infill of the cabinetry and partitions, all these provide a narrative of how the house was put together. We think of materials in terms of a layering and interweaving of relative permanence and weight. This also facilitates a stitching together of different spaces.

One of the ways in which we attempt to exploit the experience of repetition is to identify the dynamic forces or forms of a site, analyze them, and then express them as dynamic forces or forms of the architecture.

PS: The Dillon Street residence features an extreme structural stretch from the existing house to the new wall that you added....

MB: That project has a highly novel structural solution to a difficult problem. The existing structure for the house landed five feet away from a retaining wall that would not have been able to support the surcharge of any additional load. At first we ruled out any addition, as replacing the wall would have been prohibitively expensive. Then we hit on the idea of supporting a new dining room on a pylon dropping down 17 feet to the garden below. To make it look more audacious we used only a single support at one corner. The roof is cantilevered in a similar fashion. And, there is no steel in this project.

Our goal is to reexamine conventional solutions and to find different solutions for problems whose answers have become routine. We propose to undermine reliance on the idea that there is one, ideal solution to a design problem. Another dimension to our pulling functional, formal, and material elements away from each other is to attempt a kind of deconstruction of the architecture. We like to break down a building into its component parts, to pull the structure out and to show the differences between materials.

我们感兴趣把材料作为建筑制造的一种编年史来表达。在我们自己的房屋里，混凝土基墙、厚重钢筋结构、轻型的钢筋骨架、石灰墙面、家具和隔离物的填实，所有的这些使得这栋房屋情节化了。我们考虑了材料的层次和相互间的相对持久性和重量的交织，并通过它们把不同的空间连接在一起。

我们尝试去探索地形的动力或形式特点，并分析它们，以此作为建筑的表现形式。

PS：Dillon大街住宅所体现的特色就是从原有的房屋到新墙是一个结构性的极端延展……

MB：对于这个难题，这个项目创造了一个非常新颖的结构解决方案。原有的离地5英尺的结构墙不能再支撑任何附加的荷载。开始我们排除了任何附加物，因为替换这个墙面将花费昂贵。我们就有个想法，在离花园以下17英尺的门塔上支撑一个新的餐厅。为了看上去更加大胆创新，我们在一个角落只使用了一个支撑。屋顶以类似的方法悬挂起来，而且，在这个建筑中没有使用钢筋。

我们的目标是重新审视传统的解决方案，并对那些通用的惯例提出不同的解决方案。我们对设计中的问题从不采用现有的最优解决方案。我们把功能、形式和材料相互对立，以此尝试建筑的重新改造。我们喜欢把建筑分解成一个个组成部分，把结构抽出来以体现不同材料之间的区别。

The first house we designed from ground up was built on an extremely tight budget. It replaced an existing bungalow that was condemned after the 1993 earthquake. Rigorously built on a 4-foot grid, the house adapts HUD's "Value Engineered Framing" concept to its design advantage.
Adhering to the tight area of the original pad, the building diagram consists of 2 interlocking masses: a tall 24' by 24' main house volume and a 20' by 20' garage volume. The main living spaces are grouped on an open plan half of a flight up from the entry. A narrow, linear mezzanine extending from the living room to over the rear half of the garage serves as the study and a media corner. Two bedrooms are tucked into the lower level, each with double French doors opening out to the garden. A generous bathroom at the mid-level serves both floors.
The use of predictable, simple, industrial material further contributes to the house's economy, enabling it to be built very nearly on budget.
The Yale-Maclean Residence, with its modulization and palette of material, formalized a strategy for the next new houses to follow.

这是我们设计的第一栋预算极其紧张的独立住宅。它替换了一栋在1993年地震中损坏的平房。房屋严格建造在4英尺高的地基上，我们在设计当中采用了美国住房和城市发展部提倡的“价值工程框架”的观念以体现其优势。
房屋由一个24英尺×24英尺的主体房屋和一个20英尺×20英尺的车库组成，并与原始基底紧密地连接在一起。主起居空间位于距人口空间半层高的开放平面上。一个狭窄的、直线形的夹层由起居室延伸至车库的后半部，作为书房和多媒体角落。两间卧室隐藏在较低的位置上，每一间都有双扇法式门通向花园。一间巨大的浴室在楼层中间，同时为两层楼所共用。
对可预见的，简单的工业材料的使用使得房屋变得更加经济，同时也刚好没有超过其预算。
Yale-Maclean住宅以其形式和材料的结合开发，为我们建造下一栋房屋提供了新的策略模式。

1. living　2. living

1~2. 客厅

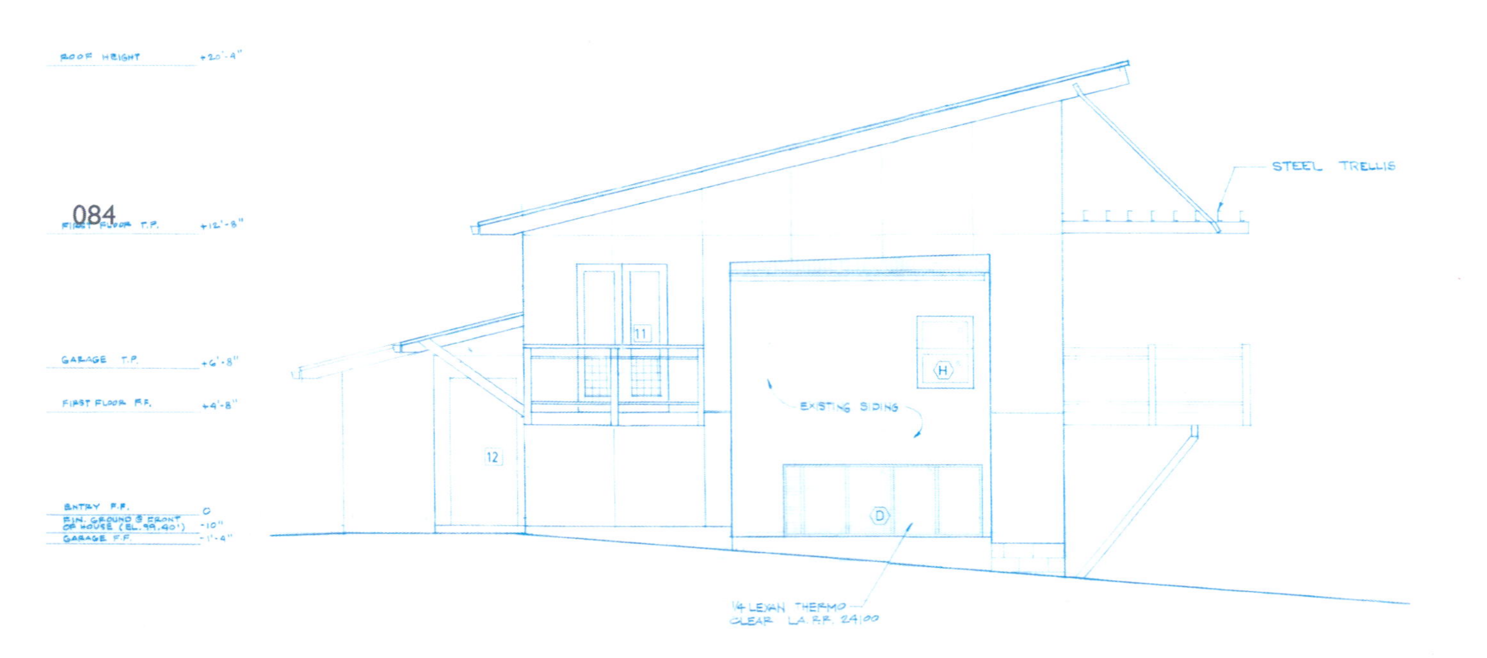

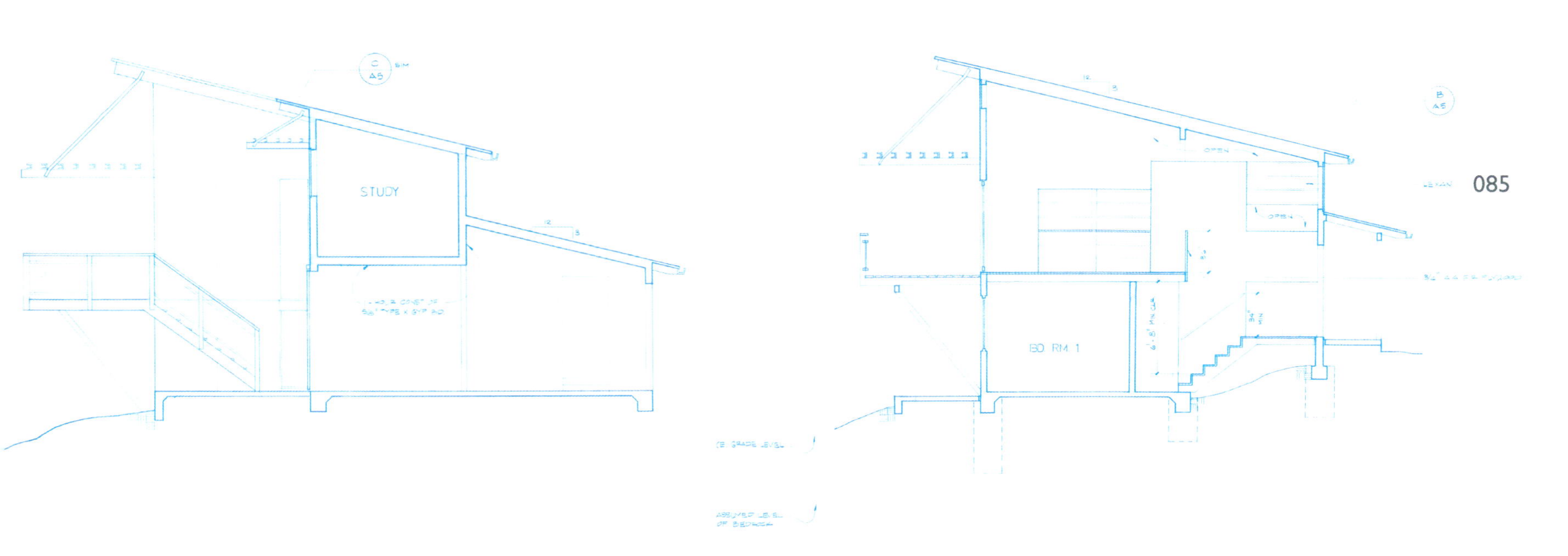
STUDY
BD RM 1

3. mezzanine　4~5. roof details　6. rear elevation

3. 夹层　4~5. 屋顶细部　6. 背立面

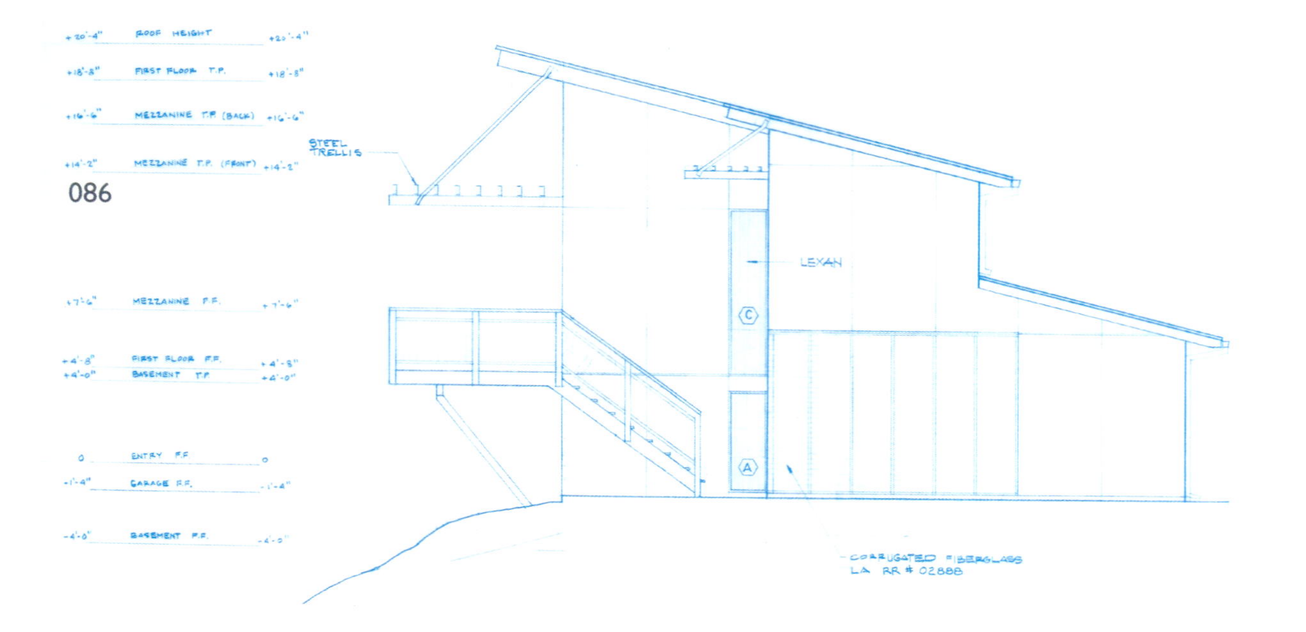

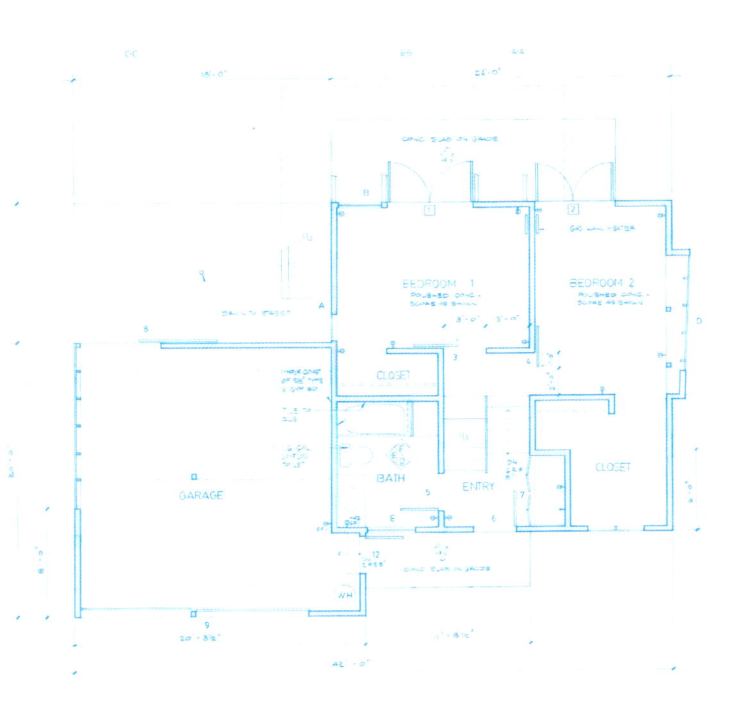
BEDROOM 1
BEDROOM 2
CLOSET
BATH
ENTRY
CLOSET
GARAGE

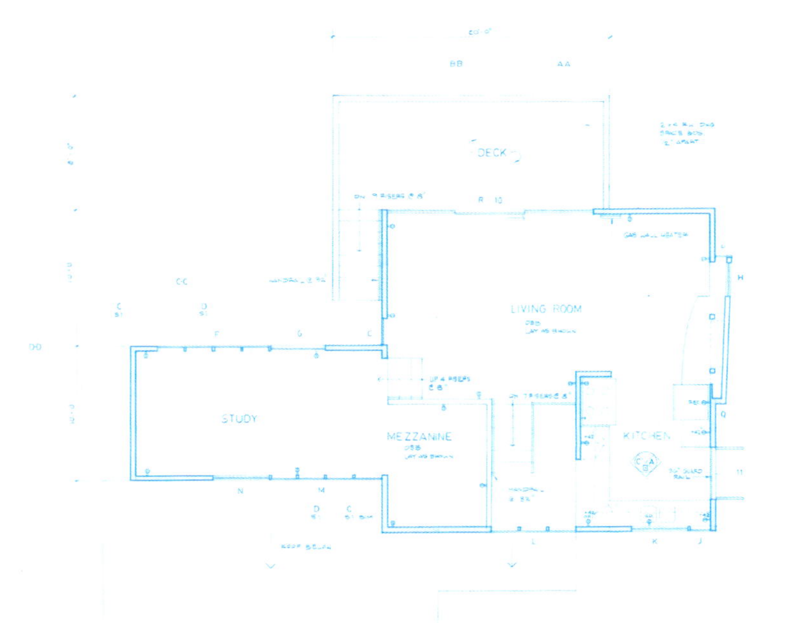
DECK
LIVING ROOM
STUDY
MEZZANINE
KITCHEN

A new 1,500 ft^2 house was built; one that perches lightly on a ridge and projects outward to command a panoramic view. An internal open stairwell with a suspended stair connects a mezzanine study, the daytime functions at mid-level, and the nighttime functions below.
Three levels of inverted gable trusses diminish the apparent weight, allowing the building to appear to float.

这栋新建的1 500平方英尺的住宅轻盈地坐落在一个山脊上并且向外投射出它的全貌。内置的悬挂式楼梯连接了上下楼层中间的书房，中间层为人们白天的活动提供了场所；楼下是人们夜间活动的所在地。
三个层次不一的倒三角形墙构架减少了外观上的负担，使得房屋看上去好像是飘浮在空中。

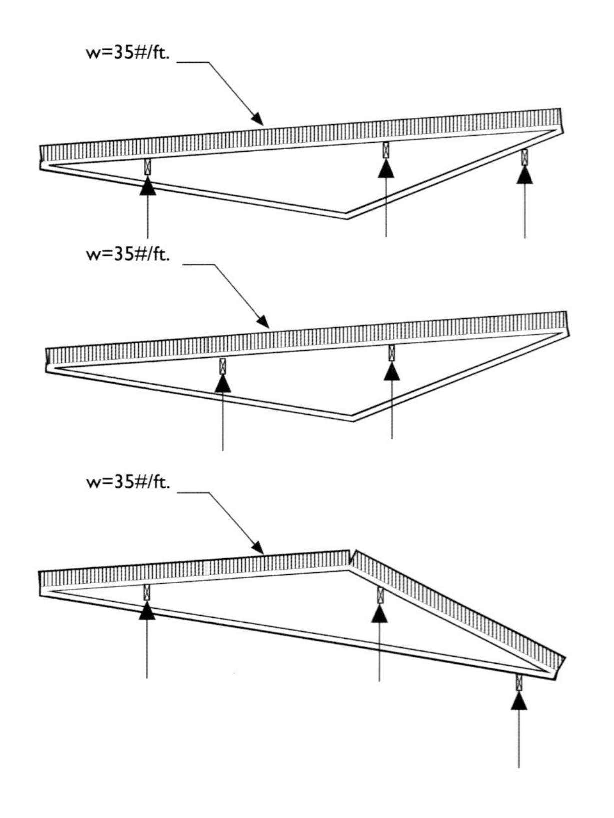

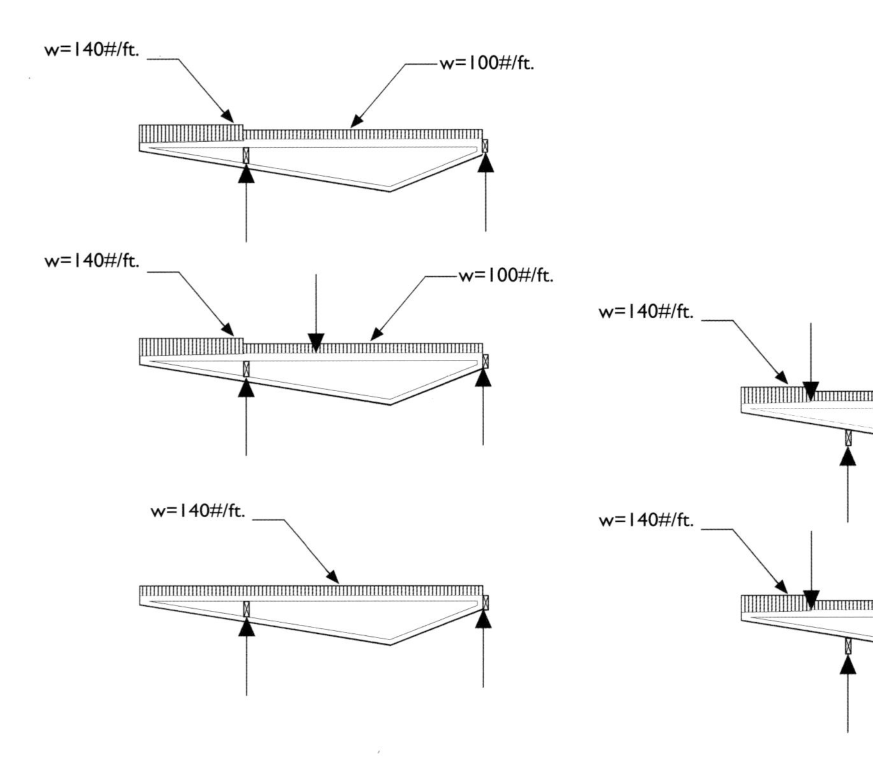

1

1. truss force diagrams 2. view from west

1. 屋架承重图解 2. 从西面看建筑

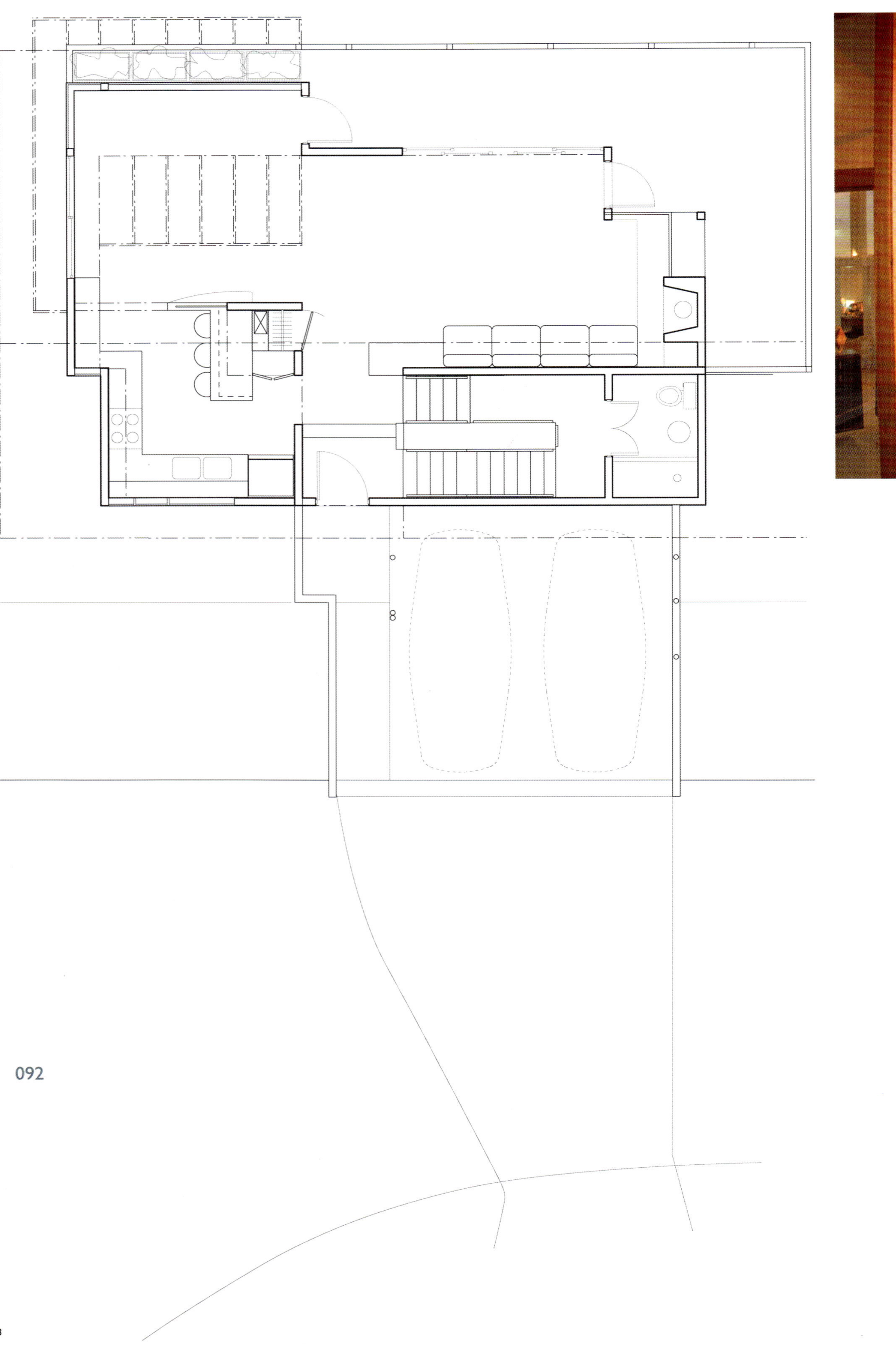

3

3. living / middle level plan 4. balcony 5. view from south

3. 起居室／中间层平面 4. 露台 5. 从南面看建筑

6. elevation from street 7. sleeping / lower level plan 8. mezzanine / upper level plan

6. 沿街立面 7. 卧室／底层平面图 8. 夹层／上层平面图

6

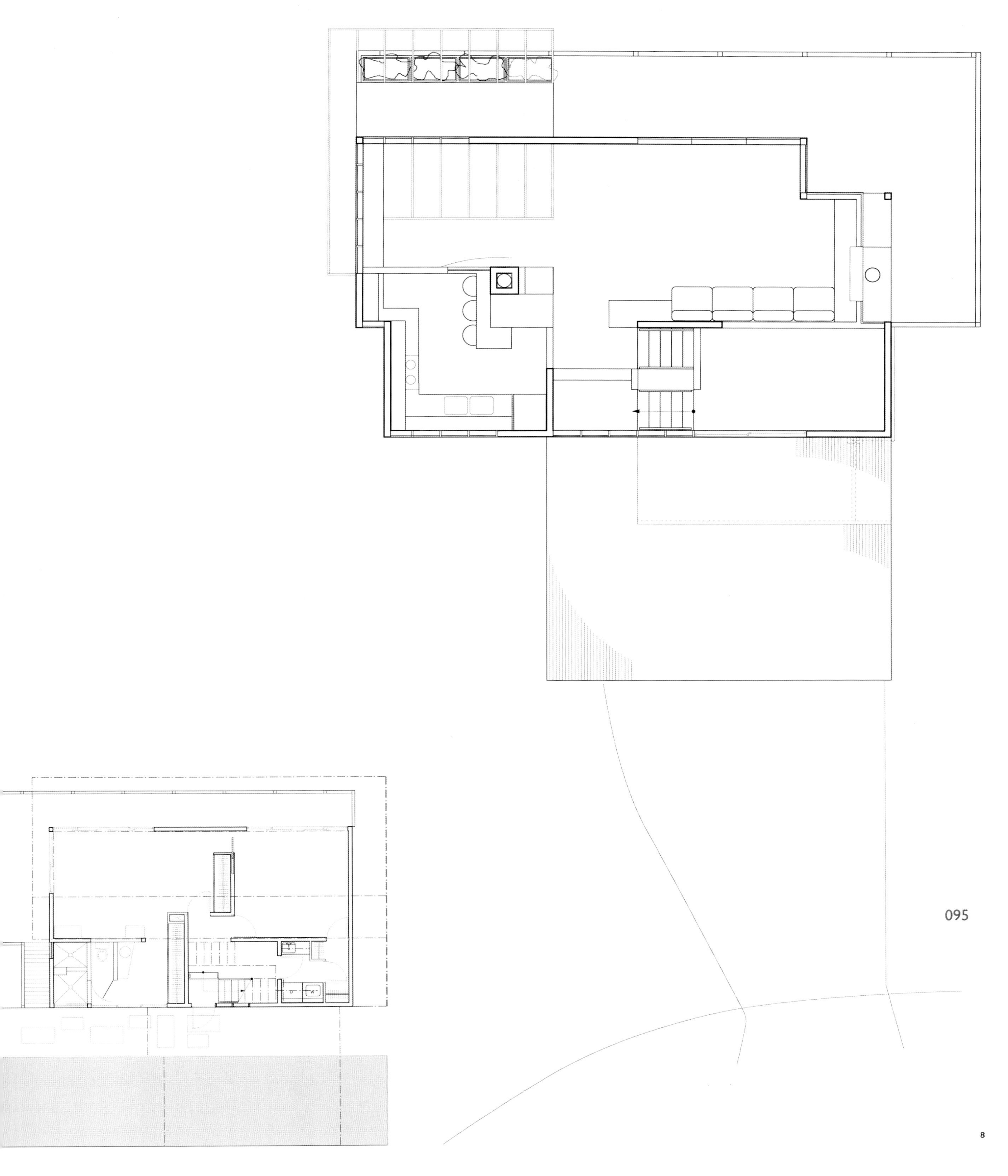

095

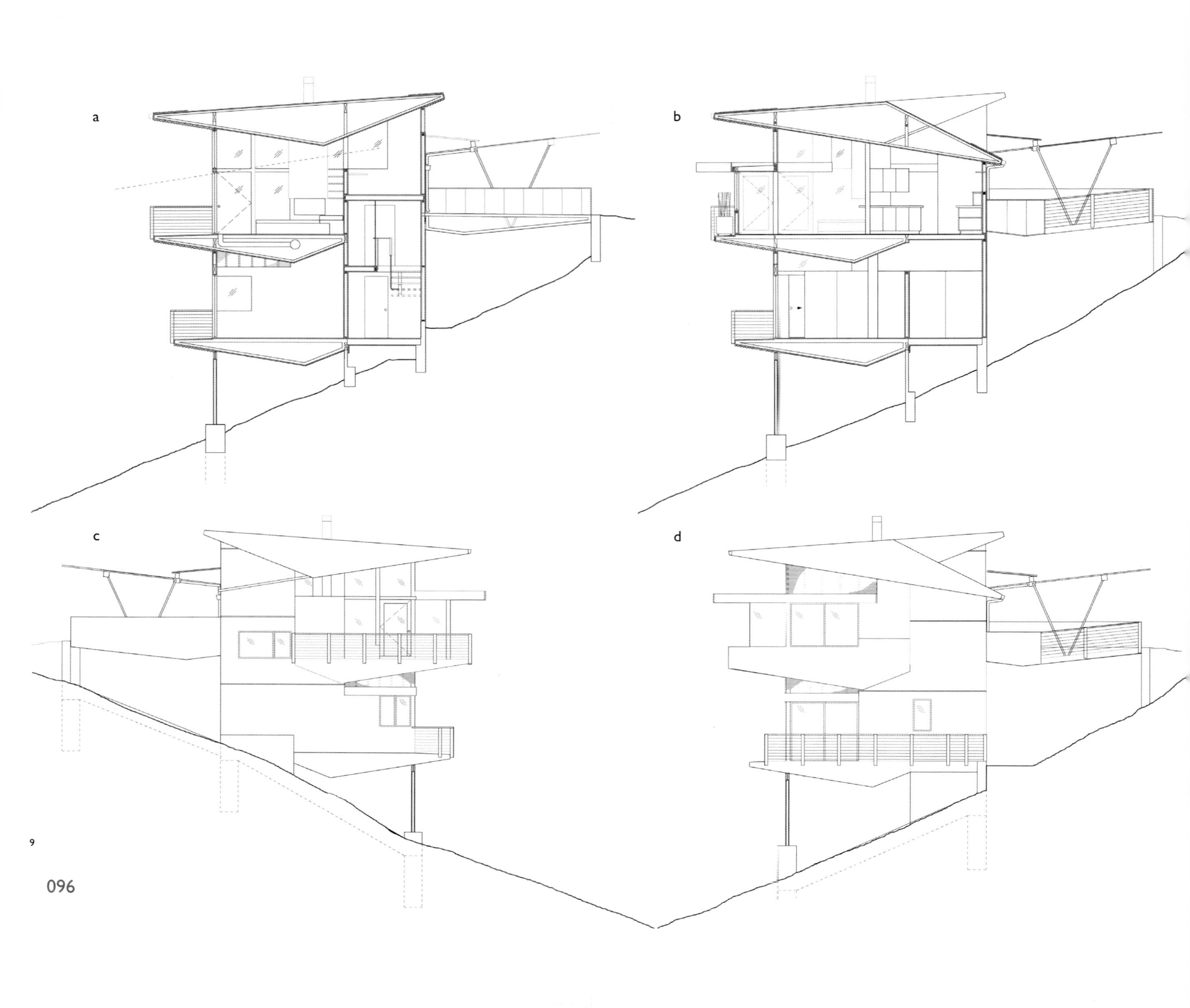

9. a: section through living looking east b: section through kitchen looking east c: west elevation d: east elevation
10. view from west 11. view from southwest
9.a 从起居室往东看剖面 9.b 从厨房往东看剖面 9.c 西立面 9.d 东立面
10. 从西面看建筑 11. 从西南面看建筑

10

11

Located in Mt. Washington, a hilly, diverse neighborhood near downtown Los Angeles; the house that we designed for ourselves rises out of a typical 50 by 100 foot urban rectangle on the north facing slope of a shallow canyon.

Built around an efficient modular armature of concrete and steel, we conceived of our own house as a flexible and expandable environment that will evolve with our changing needs. Light, spatial communication and connection to the outdoors were paramount considerations. Budget and resource conservation were practical concerns that had to be dealt with hand in hand.

To us, resource conservation begins by building small and just what is needed. At our house, outdoor spaces become extensions of interior spaces, circulation areas are pressed into double duty, and the volumes of the garage and attic were designed with future infill possibilities and flexible use in mind. In maximizing opportunities for outward extension and internal views while minimizing spatial redundancy; our house possesses a functionality, a sense of volume and openness that belie its modest 1,600ft^2.

Formally, our house is a play of volumetric contrasts. The Living quarters are located above a tall, light-filled garage volume that will accommodate a future mezzanine infill and will serve interchangeable functions. Entered through a shaded garden and a compressed foyer, the living space soars to culminate in north-facing clerestories that provide even light and passive cooling throughout the day. Tucked against the hill, the kitchen-dining space extends laterally to patios and views.

A central open stair bisects the public spaces below into living and service functions, and aligns a private study to the private spaces above. The Main Bedroom connects to an attic/loft, which will expand to function as a retreat with access to a roof terrace. Around the exterior, a living terrace on the garage roof bridges to a ground level dining patio, from which one may ascend to the upper garden and reenter the house through a common room. Stitched together by multiple inside and outside paths of circulation as it negotiates the slope via shifts of ground plane, our house enlivens our perception of what we are standing on and of where a path begins and ends.

这栋坐落于华盛顿山脉的一座小山上、靠近洛杉矶市中心的为我们自己设计的房子在小峡谷的北面斜坡上采用了典型的50英尺X100英尺的城市矩形。

房屋建立在一个经济的钢筋混凝土模架基础之上，我们将我们的房子设想成一个可以随着我们的需要的改变而改变的具有弹性和可扩张性的环境。光线、空间的联系以及与屋外的连接是首先要考虑的事情。我们也需要对预算和节约资源进行处理。

对我们而言，节约资源意味着建造小而又能刚好满足我们需要的房屋。对于我们的房子，室外空间是室内空间的延伸，流通区域被赋予双重用途，并且车库的大小和阁楼的设计都考虑到将来可能的用途和灵活利用。在尽可能将空间的浪费减至最小的同时使之具有最大的向外延伸的机会和内部景观，我们的1 600平方英尺的中等房子让人感受到大而开阔的感觉。

我们的房屋在形式上体现了不同空间面积间的鲜明对比。住宅坐落在高处，明亮的大车库可以为将来提供一个可充实的中间层，也可以在将来改作其他用途。经过一个被遮蔽住的花园和一个压缩了的休息室进入到在天窗处达到了尽头的生活区，这扇天窗为房间提供了光线和适宜的凉风。住宅的厨房和就餐区依山而建，从侧面延伸至天井和观景处。

一个开在中间的楼梯将下面的公共空间分为居住的和服务性的两部分，同时使得私人的书房和上面的私人空间连成一行。主卧连接着一个将来可以扩展为顶层阁楼的露台。在房屋外面，车库顶上的平台和一个作为户外进餐时用的天井相连，人们可以从这个天井进入到上面的花园，并可以通过一间公共的房间重新回到房屋内。室内外的多重通道依据建筑的坡度相互交会流通，这栋建筑体现了我们所认知的，也就是我们所坚持的起点和终点。

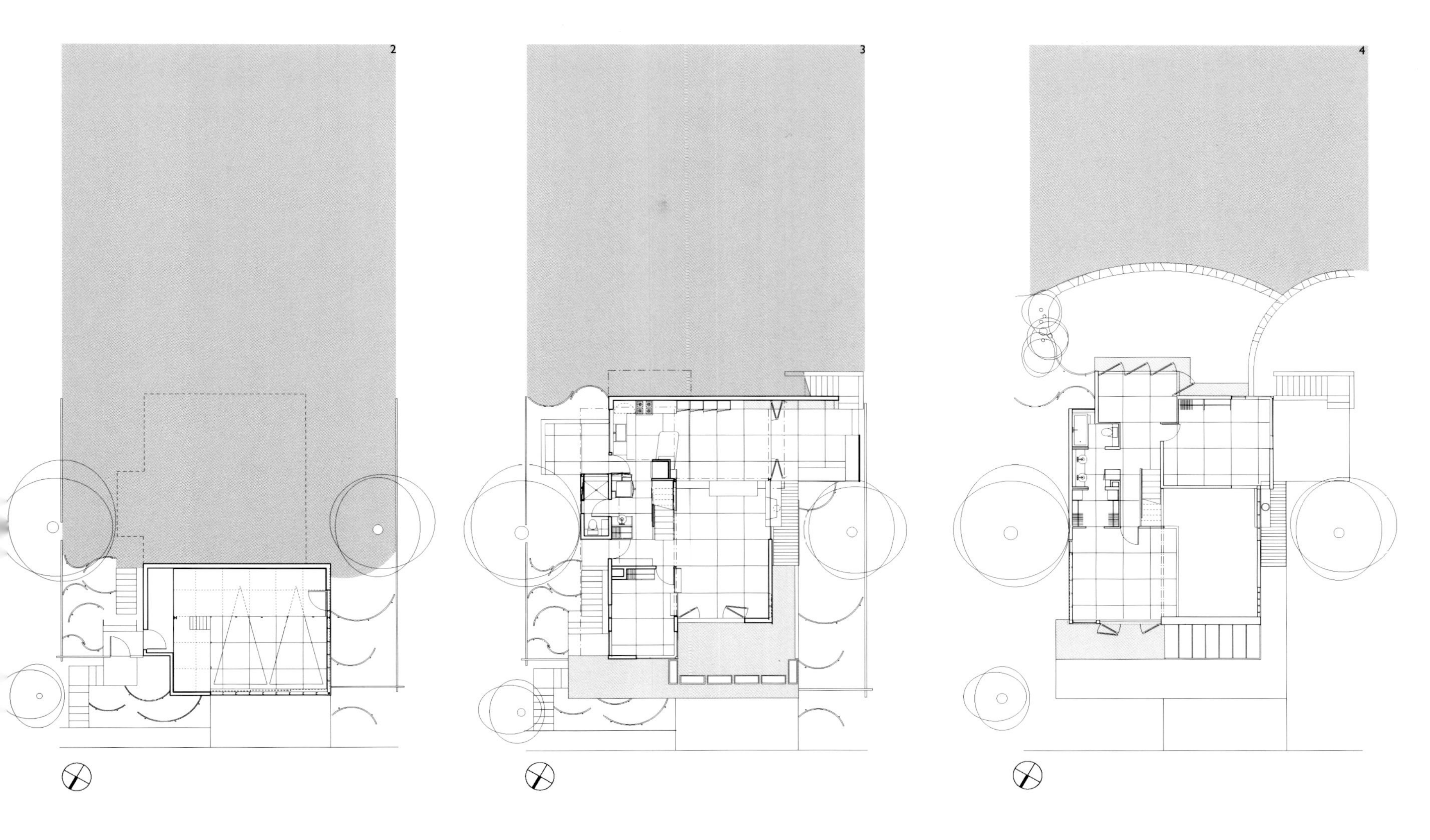

1. view from northwest　　2~4. section level plan

1. 从西北面看建筑　　2~4. 各层平面图

5. bi-fold doors open living area onto living terrace
5. 连接客厅和露台的双扇折叠门

6. fireplace
6. 壁炉

7. living - vertical view
7. 客厅

8. living-dining beyond
8. 客厅－餐厅

9

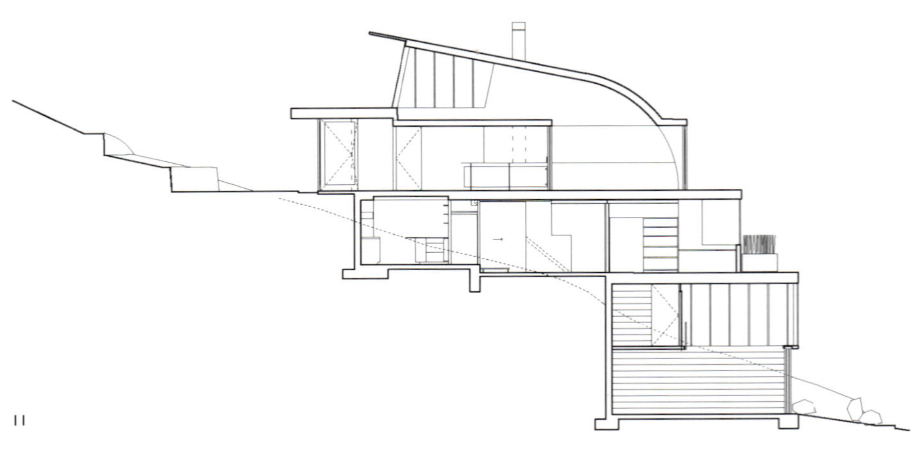

11

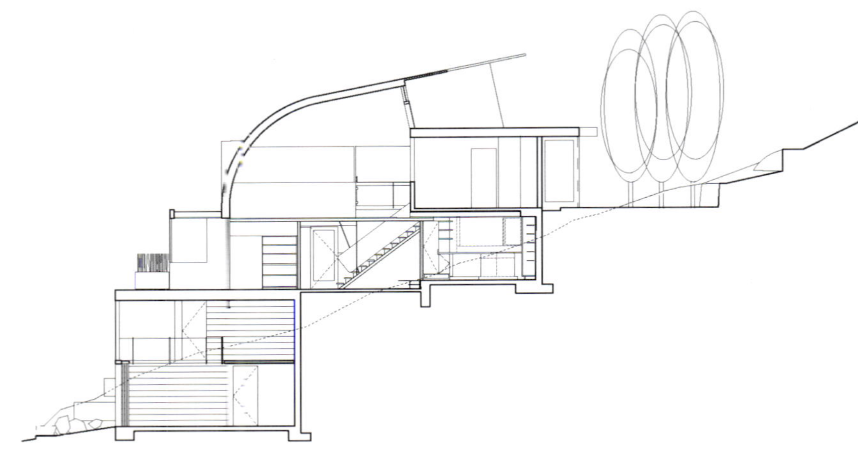

9. stair detail　10. view of stair　11. section looking west and east　12. stair

9. 楼梯细部　10. 楼梯　11. 西侧和东侧的剖面图　12. 楼梯

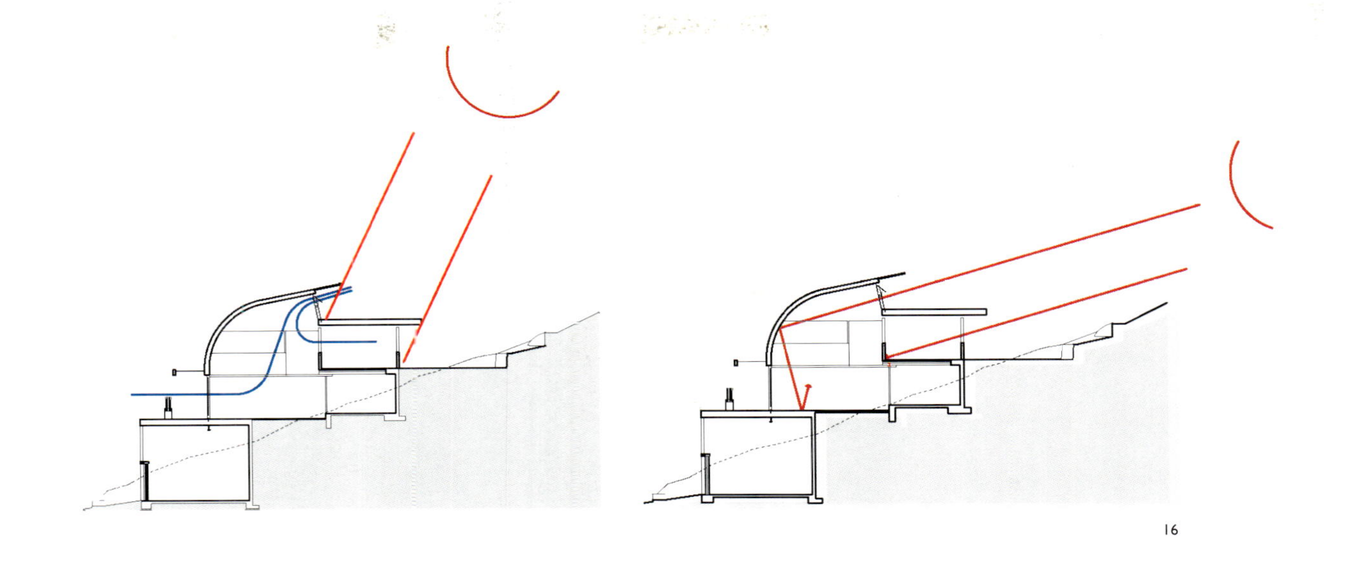

16

13. bath tile detail

13. 洗漱间瓷砖细部

14. master bedroom viewed from loft - doors open to bedroom terrace

14. 从阁楼看主卧——卧室门通向露台

15. master bedroom terrace

15. 主卧露台

16. summer ventilation, winter heat gain diagrams

16. 夏季通风和冬季供暖设计图解

17.detail at dining

17. 餐厅细部

17

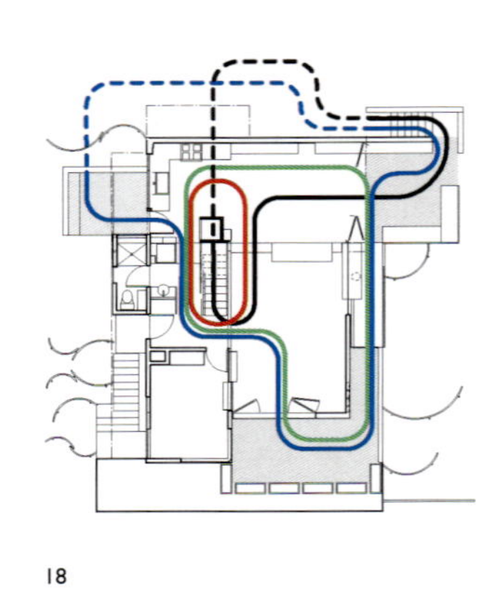

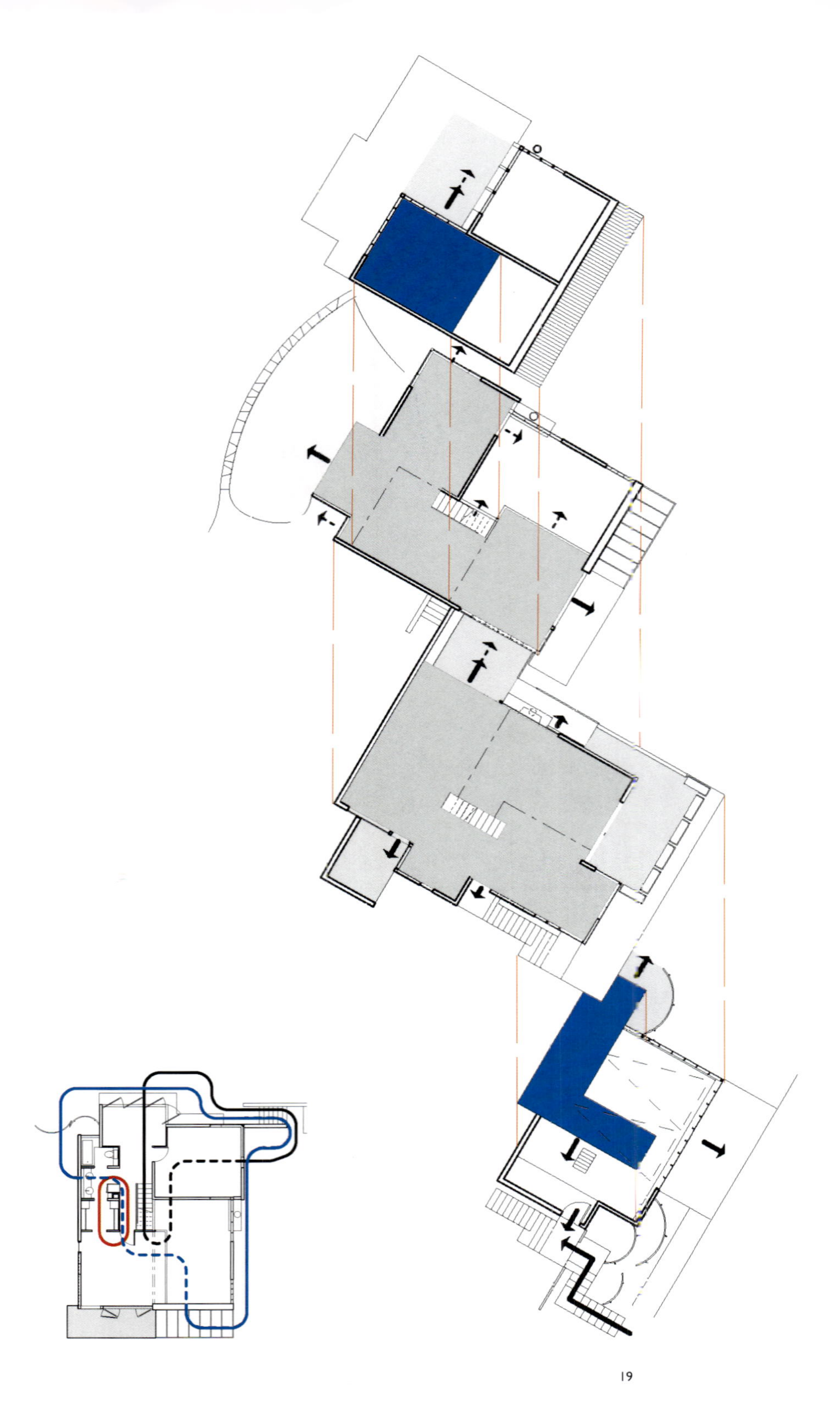

18

19

18. circulation diagrams
18. 人流交通图示

19. blue indicate future infill / arrows indicate view and access to exterior
19. 图中蓝色色块和箭头标明了建筑的外出通道

20. view from street
20. 从大街上看建筑

PS: Given that both of you began your education in fine arts, how does representation condition your design process?
MB: Ultimately, architecture is not about something; it is something. Architecture is the totally integrated experience; this is what a lot of art tries to get at. I don't want to call attention to ourselves with our buildings, I want to grab people unaware and affect them before they know someone is up to something. Movement through space, interaction with light; this is such a powerful medium. A building is the thing itself, not a model of ideas.

PS：由于你们都是从学习美术开始的，你们的设计过程是怎样体现出来的?
MB：最终，建筑不是关于某些事情，而是事情本身。建筑学是经验的完全整合，这就是许多艺术极力想达到的。我不想用我们的建筑唤来对我们自己的注意，我想在人们知道策划事情的人是谁之前不知不觉地抓住他们的注意力并影响他们。空间的移动、光线的交融，这些都是强有力的表现方法。建筑是其本身，而不是思想的模型。

PS: Well you might say that, but your work and my conversations with you indicate that ideas are entirely important to your work.
AF: Representation as a part of the design process is somewhat different from making art. In the office, we sketch, draw, and make a lot of models. We haven't had much opportunity to archive and catalog our process documents as representations of our work. We see the progression from one built project to another as achieving this same result for us. They represent a train of thought and are works in progress.

PS：你可以这么说，但你们的作品以及我与你们的谈话都表明理念在你们的作品中是十分重要的。
AF：表现方法是设计过程中的一部分，但它又不同于制造艺术。在办公室里，我们构图，绘制，制造很多模型。但是我们还没有那么多的机会像处理我们作品的设计图一样将我们在设计过程中的文档都保存并编成目录。我们看到了从一个已建成的项目到另一个项目的进步，它们对于我们取得的效果是一样的。它们代表一连串的想法，也体现了作品的进步。

The temporary nature of a gallery installation brought into focus the role that time plays in architecture. Our design for the show prioritized the real time experience that we created in the space, before the chronological presentation of our work.
Responding to the formal conditions of the 30 by 30 foot space, we resurrected and adapted a uni-strut display system long forgotten by the Gallery. We created of a kit of parts that introduced a path, moments and transitions that featured our work along the way; culminating in a place of rest with an overhead video projection of the change of light and movement through our own home.
The installation presented an armature of relationships that explored the choreographic and perceptual themes of Circularity, Shifting Points of Reference, and Reversal of Interior and Exterior.

画廊的临时性展览装置的设计重点应当是建筑的时间性表达。我们的展示设计将空间中的真实时间体验放在首位，甚至置于我们作品的年代顺序之前。
为了和30英尺×30英尺的空间条件相呼应，我们重新采用了画廊中久被遗忘的合成结构陈列系统。我们利用一系列组件创建了一个通道，代表我们作品特征的瞬间和转换被沿路引入。展示的高潮在一个顶上安置有投影仪的地段，这些视频的光线和运动的变化可以在我们自己家里控制。
这个装置表现了对编排和感知性主题之间的关系、参照物视点的控制关系，以及内部和外部互相转换等关系的探索。

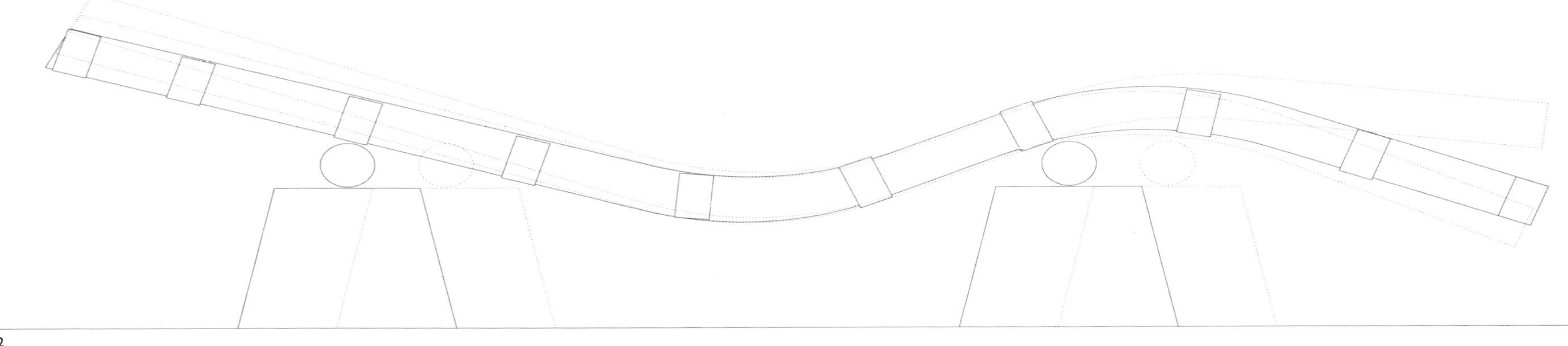

2

1. plumb bob at entrance　2. elevation of "lawn" chair　3. plan　4. looking from beginning of path to end of path beyond

1. 入口处的铅锤　2. 草椅立面　3. 总平面图　4. 从通道的起点看通道的终点

a 1st transition - beginning of path
b 2nd transition - shifting deck, connection to exterior
c observation bench
d display of work begins on walls and masonite slats
e 3rd transition - end of path
f place of rest - "lawn" chair with video projection above

a 第一层转换，通道的起点
b 第二层转换，与室外相连的控制板
c 观测台
d 从墙上和masonite百叶上开始的作品展示
e 第三层转换，通道的终结处
f 剩余空间处的草椅和顶端的投影仪

5. "lawn" chair with video projection above　6. looking in from outside　7. place of rest - "lawn" chair with video projection above　8. shifting deck, first model display

9. connection to exterior observation bench　10. place of rest - "lawn" chair with video projection above

5. 草椅和顶端的投影仪　6. 从画廊外面看室内　7. 剩余空间处的草椅和顶端的投影仪　8. 控制板和第一个模型展示　9. 与室外观测台相连的连接处　10. 剩余空间处的草椅和顶端的投影仪

10

9

Transcendence

PS: In previous conversations you alluded to the paintings of Giorgio de Chirico as influential to your work, could you expand on this?
MB: De Chirico's work depicted architecture as an emotional landscape rendered with a deeply evocative mood. My goal, when not thinking about deadlines and budgets, is to produce these kinds of transcendental, emotional spaces. There is a moment when your experience of a space overwhelms your awareness of it, when the emotional or psychological experience supersedes the intellectual.
James Turrell's light installations also are transcendental environments where the physical reality disappears into pure emotional experience. He creates a highly powerful experience based on subtle illusions. Architecture has to be very careful about the use of anything illusional, because architecture doesn't ask the participant for any suspension of belief. You don't want to undercut the persuasive reality of the thing itself.

PS：在前面的谈话中，你们提到契列柯的油画对于你们的作品有非常重要的影响，你们能详细地讲讲吗?
MB：契列柯的作品将建筑描绘成为能够引起深层情绪的情感景观。在不考虑期限和预算时，我的目标是创造出这种卓越的情感空间。总有一刻，当情感或心理的体验替代了理性时，你对于空间的体验能够淹没你对它的认识。
James Turrell的灯光装置也能营造卓越的环境，在那里物理的真实被情感的体验所消解了。他以细微的幻觉为基础创造了一种强有力的体验。在建筑中使用幻觉的东西要十分谨慎，因为建筑从不要求参与者对任何信念的放弃。你不要想去削弱事物本身的真实说服力。

AF: Turrell's detailing does not draw attention to itself but rather to the curved space. Rather than a simulacrum, Turrell transforms illusion into a pure form of transcendental perception. Gardens can do this too, with something as unspoken as the special way that light coming through an aperture has the power of making leaves into a soft pattern behind which the rest of the world disappears. Japanese and Chinese gardens tend to have these qualities that convey shape-specific states of mind. This summer, we visited the Kuppersmuhle Museum in Duisburg, Germany, designed by Herzog and de Meuron. The stairway[7], constructed of pigmented board-formed concrete (perhaps evoking barrel slats) with carefully orchestrated light sources, achieves a time essness that sweeps you into the experience of that moment. It was perhaps an example of the transcendence that the architects talked about in their book Natural History[8].

MB: We intend our architectural effects to improve the quality of life, to bring joy, to engage people emotionally and intellectually with the space around them. To the extent that these effects can be achieved transparently, that is, without calling attention to the architect, so much the better. I am interested in an architecture which does not stand apart from the surrounding pattern of life, but which rather is a subversion of it. Spaces should engage people as a dream. One should be influenced by architecture without necessarily becoming aware that this is taking place.

AF：Turrell的作品细节把对事物本身的关注引入到了弯曲空间。Turrell将幻觉转化成了一种纯粹形式上的卓越感知，而不是幻影。花园也能通过空间的方式进行无言的表达，从缝隙里透进的光线能制造一种离别后的温柔气氛，它背后多余的世界消失了。日本式和中国式的花园就趋于这些特定的情绪表达。这个夏天我们参观了德国杜伊斯堡的由赫尔佐格&德梅隆设计的Kuppersmuhle博物馆。博物馆的楼梯是由有色板形混凝土（也许是桶条）和节律感极强的光源共同建构而成，这种感觉将你带人了那一瞬间的永恒体验当中。建筑师在他们的《自然的历史》一书中谈到了这一也许是有关超越的范例 。

MB：我们试着用我们的建筑影响力来改善生活的品质，带给人们欢乐，引发人们从身边的空间中获得情感和理性。在这个意义上这些影响力都明显地达到了，确切地说，这比唤起对建筑师的注意更好。我认为建筑不能对立于它周围的生活模式，但相反地是要改变生活。空间应该带给人们梦想，建筑应该在人们并没有意识到的情况下影响他们。

Kenner Studio

A film studio cum garden room for a documentary filmmaker and his family.
The building's form responds to the concerns of scale and relationship to an established garden and existing historic main residence. The interior addresses the programmatic need for compartmentalization, sound and light control on the one hand, and the desire for openness and visual connectivity on the other.
Our strategy articulates the readings of skin versus structure, envelop versus infill; such that interiors spaces fit within building shell like drawers in a cabinet.
The birch-ply ceiling of the lower floor slides to reveal a light soffit and becomes the floor and wall of an office above. A gabled ceiling bridges over office spaces separated only by cabinetry and glass. Ipe wood slates over the downstairs Sitting Room evoke the feeling of a garden lath house, and plays with the notions of tenuous enclosure and voyeurism.

这栋花园工作室住宅是为一位纪录片制片人和他的家庭所设计的。
建筑在形式上充分考虑了比例以及原有花园和现存的历史性与住宅之间的对应关系。建筑的室内布局一方面强调了制作节目时隔声及光线控制的需要，另一方面也满足了对开阔的视觉联系的要求。
建筑的内外结构清晰明了。其内部空间布局置于整个建筑之内恰如其分，就像镶嵌在橱柜里的抽屉一样。
底层由桦树木制成的顶棚延伸出一个轻型的拱腹，并形成了上层办公室的地板和墙壁。跨越整个办公室空间的三角形挑梁被立柜和玻璃隔开。楼下起居室被木排围了起来，使得整个房间具有一种木栅栏花园的感觉，这里运用了低密度封闭和窥蔽主义的观念。

1. garden room exterior at dusk 2. studio entrance

1. 黄昏中的花房外景色 2. 工作室入口处

3~4. office window and office　5. stair landing

3~4. 办公室的窗户和室内　5. 楼梯平台

6

6. view from top of stair

6. 从楼梯上端往下看

7. view of hallway from main work space, sliding panels (in foreground) convert area into editing room

7. 从工作间看出去的室内走廊，可滑动的隔断可以将这一区域转换为电影剪辑室

7

8. view from garden　9. garden room

8. 从花园看建筑　9. 花房

1.Vine Street Studio, Hollywood CA, completed 1997 2.Maunu Residence, Mt. Washington, CA completed 1996 3.Ohayon-Van de Sande Residence Los Angeles, CA, completed 1998 4.Martin Studio No. Hollywood CA completed 1999 5.Campbell Apartment Venice CA, completed 2001 6.Rhodes Residence Echo Park CA completed 1996 7. Katz-Simmons Residence, Lake Sherwood completed 1999 8. Berman Apartment Santa Monica CA, completed 1991 9. Hunt Design Studio Pasadena CA (unbuilt) 10. Chesser Residence Los Feliz CA, completed 1991 11. Johnson-Rose Residence Hollywood CA (unbuilt) 12. Lindley-Richardson Residence Pacific Palisades, CA, completed 1992 13. Hillside Residence Mill Valley, CA completed 1996

Fung+Blatt 的其他已完成的住宅项目

Brick Pillow by Ali Acerol

Ali Acerol 制作的枕砖

Kuppersmuhle Museum stairway by Herzog & de Meuron

赫尔佐格 & 德梅隆设计的 Kuppermunle 博物馆的楼梯

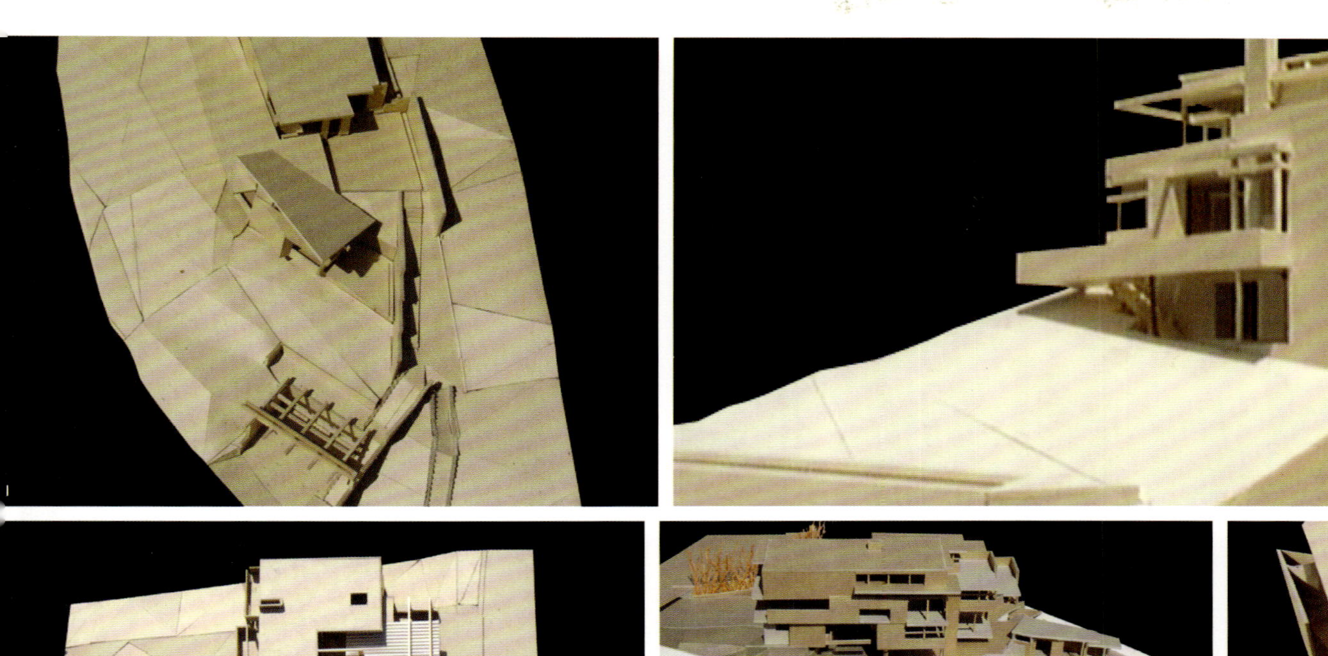

1. Leach Residence, Mt. Washington, CA / to be completed 2004 2. Albee Residence, Mt. Washington, CA / to be completed 2005 3. Noonan Residence, Mt. Washington, CA / to be completed 2004

4. Lefevre Residence, Mt. Washington, CA / to be completed 2006 5. Katz - Simmons Residence, Tarzana, CA / to be completed 2005

Fung+Blatt 的正在建造中的住宅项目

Books + places

PS: What are you reading these days?
AF: The last memorable books I read that related to architecture were Norman Klein's The History of Forgetting: Los Angeles and the Erasure of Memory and Herzog & de Meuron: Natural History, edited by Philip Ursprung. Klein wrote about the fabrication of collective memory where one has been erased. Ursprung explored the place of cultural memory in Herzog and de Meuron's architecture. I've also devoured some novels and essays by East Indian writers in the past year -- Death of Vishnu, Interpreter of Maladies, The God of Small Things and some political essays by Arundhati Roy.
My interest in Indian writers stems from my childhood memory of living in the colonial environment of Hong Kong. I lived next door to a third generation Indian family. The grandfather was recruited to Hong Kong by the English as a wrestler. How people end up where they are is fascinating to me. I would like to trace these invisible paths, and look at the patterns that emerge, not just from the social-historical point of view but also from that of an artist.
On the other hand, India epitomizes the other end of globalization. The contrasts that exist there and in many other emerging countries are disturbing and surreal. The seeming urgency, the momentum toward the erasure of memory, the balance tipping toward inadvertent destruction is a serious concern for anyone who has brought children into the world.

PS：你们目前都读一些什么书？
AF：在我最近读的关于建筑的书中，印象最深的是Norman Klein的《遗忘的历史：洛杉矶和被抹掉的记忆》和由Philip Ursprung编辑的《赫尔佐格&德梅隆：自然的历史》。Kline写的是关于集体记忆的形成，在那里个人的记忆已经被忽略。Ursprung探索了赫尔佐格&德梅隆的建筑中的文化记忆。过去几年里，我还囫囵吞枣地读了一些东印度作家写的小说和散文——《毗瑟拏之死》，《道德的缺失》，《小事之神》和一些由Arundhati Roy写的政治短文。
我对印度作家的兴趣主要来自我童年时香港殖民地生活环境的回忆。我隔壁住着一个三代同堂的印度家庭，爷爷是由英国人征募到香港的摔跤手。人们在他们所在地的生活终结深深地吸引着我。我想追寻这些无形的通道，并且不仅从社会、历史的角度，而且也从一个艺术家的角度着眼于其显现形成的模式。
另一方面，印度也是全球化进程完结的一个缩影。那儿存在着与其他发展中国家相类似的混乱和不真实。表面的急迫，记忆的丧失，这些在发展中造成的不可避免的毁坏是每一个将孩子带向世界的人所应当认真思考的。

PS: What places inspire your work?
MB: Other than the places that we have spoken about, I am drawn to places of extreme contrasts: desert oases, for example. Hong Kong is also such a place. The sleek modern high-rises butt right up against the rugged wilds of the peak, which towers over them. Places with extreme changes in vertical level also fascinate me. The old city of Salzburg where a tunnel cuts through the cliff face like a door in a wall, while a freestanding elevator nearby shaves half a mile off the journey to the summit. There is a quality about these places that disrupts your perceptual reference system and frees you to develop new ways of relating to the environment.
AF: I am inspired by many places in nature. I am drawn to Death Valley, California, because some of the world's oldest geology collides with some of its youngest there, and you can read it, plain as day. Portions of California's central coast, in the same way, remind me that the land is a living organism that is fragile and changeable. The Serpent Mound, a massive earthwork created by the Native Americans in Ohio centuries ago, is one of the most spiritually powerful and mysterious places I've been to. I've been drawn to Chinese gardens where sequential experiences and framing of perspectives are choreographic devices that we use in our work. I have deeply etched memories of the environment of my childhood, which I'm sure somehow affects the places that we create. I think that as we have matured as architects, we have become less impressed by the form-driven and more drawn to the contemplative and the transcendent.

PS：什么地方激发了你的工作灵感?
MB：与我们前面所提到的地方不同，我想说的是与之完全相反的那些地方：就像是沙漠绿洲，香港就是这样一个地方。豪华的现代化高楼大厦与粗旷的山野遥相呼应。一些有急剧变化的地方也同样使我着迷。在萨尔茨堡老城，从峭壁上穿过的隧道就像墙上的门一样，同时一个独立的升降机立在离顶峰半英里外的峭壁旁。这些地方有一种能破坏掉你的知觉参照系的品质，使你能够自由地创造出与环境相关联的新方法。
AF：我从自然界中的许多地方获得灵感。我想说加利福尼亚的死亡之谷，因为一些世界上最古老的和最年轻的地质学冲突都发生在那里，你可以读懂它，并可以感受到时光的威力。同样，加利福尼亚中心海岸的一部分在警示我们——陆地是脆弱而且可变的生命有机体。俄亥俄州的Serpent Mound是几个世纪前由美洲土著建造的巨大土木工事，这是我到过的最有精神力量和最神秘的地方之一。 我也一直从中国园林中吸取连续体验和透视取景这些传统设计编排法。我儿时的生活环境深深地刻在了我的脑海中，我确信它对我的创作有巨大的影响。我想作为一个成熟的建筑师，我们会变得更少为形式所累，更多的去思考和超越。

1 Arts and Architecture, John D. Entenza (John D. Entenza 1938-1962).

2 The Studio Book, Melba Levick Photography, Kathleen Riquelme text (New York: Universe/Rizzoli, 2003), "North Light and a Pool: the Hubert Schmalix Studio".

3 Privacy and Publicity, p. 234, Beatriz Colomina (MIT Press, 1994).

4 Casa Malaparte, Marida Talamona (Princeton Architectural Press, 1992).

5 Casa Malaparte, Karl Lagerfeld, Eric Pfrunder, Gerhard Steidl (Hatje Cantz Publishing 1999).

6

Brick Pillow by Ali Acerol

7

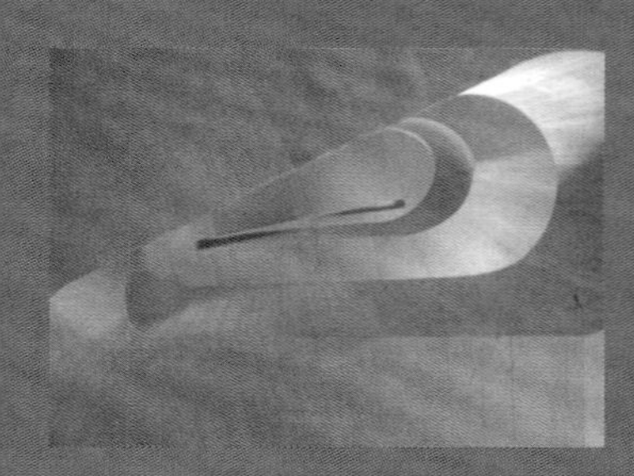

Kuppersmuhle Museum stairway by Herzog & de Meuron

8 Herzog & de Meuron: Natural History, Philip Ursprung (Lars Muller Publishers, 2003).

Chronology

Born and raised in Los Angeles, Michael Rosner Blatt studied studio art at the University of California at Santa Cruz before receiving his Master of Architecture degree from the Southern California Institute of Architecture.

Although keenly interested in architecture as a child, Blatt put off any formal study of the field for as long as possible in order to avoid the possibility of entering a career. His application to graduate school was only made upon the realization that the alternative would be to look for a job.

While attending architecture school, Blatt continued to think of himself primarily as an artist, and did not know to what use he might put his education. Nonetheless, upon graduating, he was eager to try out his newly acquired skills, and eagerly accepted a commission from his father to design and construct a painting studio.

While constructing the studio, he took a part-time position at the engineering office of Gordon Polon. Although hired as a draftsman, he was quickly put to work engineering houses. Although having no formal training in the field, Blatt studied independently with a particular focus on how engineering principals and a spirit of experimentation contributed to California Modernism. He continued at the firm for three years, where working with numerous architects gave him a foundation in the practical aspects of design. Blatt has continued to engineer most of his firm's work to date.

Alice Fung was born in Hong Kong and came to the United States as a high school student.

She received her Bachelor of Arts degree from Oberlin College in Ohio followed by a Master of Architecture degree from the Graduate School of Architecture and Urban Planning from UCLA.

Fung's interest in the built environment began in her first fifteen years in Hong Kong, where she witnessed a city struggling to keep up with unprecedented growth. In college she studied studio art / art history and urban studies, and in her last year, discovered modern dance as a medium of experiencing time and space. She chose to pursue architecture as a discipline that combines her earlier interests. In graduate school, she studied with Frank Israel, Charles Moore, Dolores Hayden, and interned with the landscape architecture firm of Campbell & Campbell. These experiences began her understanding of architecture as a continuum of the larger environment.

After graduate school, while working for Frederick Fisher Architects, Michael Blatt came in to engineer a project. They met and collaborated on a project on their own which received an award from the American Institute of Architects and Sunset Magazine. In 1990, at the beginning of an economic recession, they married and open their own studio.

Fung has taught architecture at Pasadena City College, Woodbury University and is currently an Adjunct Professor at Art Center College of Design.

They live with their daughters, Kai and TÈa in Los Angeles.

lice Fung

961

orn in Hong Kong

976

rrived in United States

DUCATION

983

achelor of Arts

)berlin College, Oberlin, Ohio

lajor Studies: Studio Art & Art History, Urban

tudies

986

laster of Architecture

raduate School of Architecture and Urban

lanning

Jniversity of California, Los Angeles, California

ROFESSIONAL EXPERIENCE

icensed Architect, State of California

988 - present

UNG + BLATT ARCHITECTS, Partner

994 -1995

EVIN & ASSOCIATES, Associate Architect

986 -1988

REDERICK FISHER ARCHITECT, Design

Associate

1985 -1986

GORDON POLON & CO., Consulting

Engineers,

Production Assistant

FRANK O. GEHRY & ASSOCIATES,

Presentation Assistant

1984-1985

CAMPBELL & CAMPBELL, Architects +

Landscape

Architects, Production Assistant

TEACHING EXPERIENCE

2002 - present

Adjunct Faculty, ART CENTER COLLEGE OF

DESIGN

Instructed and developed curriculum for

interdisciplinary

design studios

2000 - 2002

Adjunct Faculty, WOODBURY UNIVERSITY

Instructed and developed curriculum for 2nd &

3rd year architectural design studios

1998 - 2001

Adjunct Faculty, PASADENA CITY COLLEGE

Instructed and developed curriculum for studio

& lecture

courses on architectural design

Michael Rosner Blatt

1960

Born and raised in Los Angeles, California

EDUCATION

1985

Master of Architecture

Southern California Institute of Architecture, Los

Angeles, California

1983

Foreign Study

Architecture Program at Vico Morcote,

Switzerland

1982

Bachelor of Arts

University of California, Santa Cruz, California

Major Studies: Studio Art

PROFESSIONAL EXPERIENCE

Licensed Architect, State of California

1988 - present

FUNG + BLATT ARCHITECTS, Partner

1985 -1988

GORDON POLON & CO., Consulting

Engineers, Structural

Designer

1985

RUBEN OJEDA ARCHITECT, Design Associate

List of work

1988

Jonathan Blatt Studio, Los Angeles

1990

Dillon Street Residence, addition + remodel,

Silverlake

Alexander-Layton Residence, remodel,

Hollywood Hills

Berman Apartment, Santa Monica

Rolle Street Studio Duplex, unbuilt, El Sereno

1991

Maunu Residence, 2-story outdoor living space,

Mt Washington

Grimes Residence, 2nd story addition +

remodel, Mt. Washington

Rhodes Residence, remodel, Echo Park

Chesser Residence, remodel, Hollywood

1993

Draper-Wolff Residence, remodel, Studio City

Lindley-Richardson, 2nd story addition +

remodel,

Pacific Palisades

Parodi-Fair Recording Studio, Studio City

1994

J & R Blatt Residence, 2nd story addition +

remodel, Los Angeles

Costigan Residence, 2nd story addition + remodel, Ventura
Alexander-Layton Residence, entry court + outdoor living space, Hollywood Hills

1995
Hillside Residence, addition + remodel, Mill Valley, Northern California
Johnson-Rose Residence, remodel + hardscape, Hollywood Hills
Maunu Residence, Exterior Remodel, Mt Washington
Yale-Maclean Residence, Echo Park

1996
Chrysanthus-Becker Residence, 2nd story addition + remodel, Hollywood Hills
Martin Studio, North Hollywood
Mind's Eye Headquarters, Novato, Northern California
Letz Residence, 2nd story addition + remodel, Berkeley, Northern California

1997
Maunu Studio, Mt. Washington
McCarthy Residence, addition, Pasadena
Vine Street Studios, landscape + interiors, Hollywood
Katz-Simmons Residence, restoration + remodel of a Neutra house, Lake Sherwood

1998
Chang-Sklaroff Residence, Palo Alto, Northern California
Leonard Office, Pacific Palisades
Shelgren Guest House, Pasadena

1999
Barnhouse Studio, Los Angeles
Ohayon-Van de Sande Residence 2nd story addition + remodel, Hollywood Hills
Cirque du Soleil Corporate Residence, (unbuilt), Beverly Hills

2000
Schmalix Residence, Mt. Washington
Walecka-Hunt Residence, remodel, Mt. Washington
Hunt Design Studio, office interior (unbuilt), Pasadena
Miyoshi Residence, Mt. Washington
Public Storage Remediation, Glassell Park

2001
Nagler Residence, Manhattan Beach
Maunu Residence, remodel, Altadena

2002
Zempel Residence Remodel, Mt Washington
Sterritt Studio, Altadena
Strick Residence, interior remodel, Westwood

2003
Anderson Residence, Mill Valley, Northern California
Campbell Apartment, Venice
Fung + Blatt Residence, Mt. Washington
Zocalo Restaurant, Alhambra
Shift Exhibition, LAHC Fine Arts Gallery, exhibit design + installation, Wilmington

2004
Kenner Studio, Hancock Park
Maunu Pool house, Altadena

Current Projects (in Construction)

Johannessen Residence, 2nd story addition + remodel, Pacific Palisades
Leach Residence, addition, Mt. Washington
Noonan Residence, Mt. Washington
Pray Residence, addition, Mt. Washington
Katz-Simmons Residence, restoration + remodel of a house by Irving Gill, Tarzana

Current Projects (in design)

Urban Village, transit oriented mixed-use development, Highland Park
Costigan Residence, 2nd story addition + remodel, Ventura
Lefevre Residence, Mt. Washington
Albee Residence, Mt. Washington
Ames Residence, remodel, Altadena
Fong Residence, remodel, Mt. Washington
Letz Residence, addition + remodel, Stinson Beach, Northern California
Katz-Simmons Poolhouse, Tarzana
Echo Street Project, 26 unit live-work development, Highland Park
Wright Residence, remodel + hardscape, Mt. Washington
Hlidal Residence, remodel, Hollywood Hills
Haslinger Studio, Van Nuys
Goldstone Studio, artist-in-residence development, Los Angeles
Schuman Residence, Mt. Washington

Note: unless otherwise indicated, all of the above projects are located in Southern California

Publications and Presentations

Publications

Houses, Loft Publications, Spain 2004

Inside Out,, Australia

Living Etc., England, " " 10.2004

Dwell, "Domestic Democracy" 9.2004

Los Angeles Magazine, "Shady character" 7.2004

Brave New Houses: The Best New Houses of Southern California, Rizzoli Press 2003

Studio Style, Rizzoli Press 2003

The Bath Handbook, Taunton Press 2004

Conran, Liveit, England, "Art House" 2002

Wohntraum, Germany, "L.A. Story" 1.2002

Baumeister, Germany, Cover article "Lessons from California" 11.2001

Los Angeles Times Magazine, "Steeling Beauty" 6.2001

Building in Los Angeles, A SCI-ARC Publication 2001 & 1997

Residential Architect, "Mid-century Makeover" 10.2000

American Home Style, "Loft Style" 10/11.1997

Sunset Magazine, " Living on the Edge" 9.1996

Metropolitan Home, "A Colorful Renovation That is Rich in Spirit" 1/2.1995

Sunset Magazine, " A Remodel That Respects L.A.'s Tradition of Modernism" 10.1991

Presentations + Exhibits

SHIFT, An Insallation, 2003
LAHC Fine Arts Gallery, Willmington, CA

Wedge Gallery, 2001
Woodbury University, Burbank, CA

Volume 5, "Freeforum", 2000
Los Angeles, CA

Alumni Exhibit, 2000
SCI-ARC, Los Angeles, CA

Los Angeles Forum for Architecture and Urban Design, Lecture Series, 1999
Los Angeles, CA

Awards

Reader's Home Award, Metropolitan Home 1995

AIA and Sunset Magazine Western Homes Award 1991-1992

图书在版编目(CIP)数据

冯+布拉特/蓝青主编，美国亚洲艺术与设计协作联盟(AADCU).
北京：中国建筑工业出版社，2005
(美国当代著名建筑设计师工作室报告)
ISBN 7-112-07400-2
Ⅰ.冯… Ⅱ.蓝… Ⅲ.建筑设计－作品集－美国－现代 Ⅳ.TU206
中国版本图书馆CIP数据核字(2005)第043696号

责任编辑：张建　黄居正

美国当代著名建筑设计师工作室报告
冯+布拉特

美国亚洲艺术与设计协作联盟(AADCU)
蓝青　主编
*
中国建筑工业出版社 出版、发行(北京西郊百万庄)
新 华 书 店 经 销
北京华联印刷有限公司印刷
*
开本：880×1230毫米　1/12　印张：12½
2005年8月第一版　　2005年8月第一次印刷
定价: **125.00** 元
ISBN 7-112-07400-2
(13354)

(邮政编码　100037)
本社网址：http://www.china-abp.com.cn
网上书店：http://www.china-building.com.cn